Festival Places

TOURISM AND CULTURAL CHANGE
Series Editors: Professor Mike Robinson, *Centre for Tourism and Cultural Change, Leeds Metropolitan University, Leeds, UK* and Dr Alison Phipps, *University of Glasgow, Scotland, UK*

Understanding tourism's relationships with culture(s) and vice versa, is of ever-increasing significance in a globalising world. This series will critically examine the dynamic inter-relationships between tourism and culture(s). Theoretical explorations, research-informed analyses, and detailed historical reviews from a variety of disciplinary perspectives are invited to consider such relationships.

Full details of all the books in this series and of all our other publications can be found on http://www.channelviewpublications.com, or by writing to Channel View Publications, St Nicholas House, 31–34 High Street, Bristol BS1 2AW, UK.

TOURISM AND CULTURAL CHANGE
Series Editors: Professor Mike Robinson, *Centre for Tourism and Cultural Change, Leeds Metropolitan University, Leeds, UK* and Dr Alison Phipps, *University of Glasgow, Scotland, UK*

Festival Places
Revitalising Rural Australia

Edited by
Chris Gibson and John Connell

CHANNEL VIEW PUBLICATIONS
Bristol • Buffalo • Toronto

Library of Congress Cataloging in Publication Data
A catalog record for this book is available from the Library of Congress.
Festival Places: Revitalising Rural Australia/Edited by Chris Gibson and John Connell.
Tourism and Cultural Change
Includes bibliographical references and index.
1. Festivals--Australia. 2. Festivals--Australia--Case studies. 3. Country life--Australia.
4. Australia--Social life and customs. I. Gibson, Chris, 1973- II. Connell, John,
1946- III. Title. IV. Series.
GT4890.F48 2011
394.230994–dc22 2010041493

British Library Cataloguing in Publication Data
A catalogue entry for this book is available from the British Library.

ISBN-13: 978-1-84541-167-1 (hbk)
ISBN-13: 978-1-84541-166-4 (pbk)

Channel View Publications
UK: St Nicholas House, 31–34 High Street, Bristol BS1 2AW, UK.
USA: UTP, 2250 Military Road, Tonawanda, NY 14150, USA.
Canada: UTP, 5201 Dufferin Street, North York, Ontario M3H 5T8, Canada.

Copyright © 2011 Chris Gibson, John Connell and the authors of individual chapters.

All rights reserved. No part of this work may be reproduced in any form or by any means without permission in writing from the publisher.

The policy of Multilingual Matters/Channel View Publications is to use papers that are natural, renewable and recyclable products, made from wood grown in sustainable forests. In the manufacturing process of our books, and to further support our policy, preference is given to printers that have FSC and PEFC Chain of Custody certification. The FSC and/or PEFC logos will appear on those books where full certification has been granted to the printer concerned.

Typeset by Techset Composition Ltd., Salisbury, UK.
Printed and bound in Great Britain by Short Run Press Ltd.

Contents

List of Figures and Tables vii
Contributors xi
Preface xv

Part 1: Exploring Rural Festivals

1 The Extent and Significance of Rural Festivals 3
   C. Gibson, J. Connell, G. Waitt and J. Walmsley

2 Histories of Agricultural Shows and Rural Festivals in Australia .. 25
   K. Darian-Smith

3 Rural Festivals and Processes of Belonging 44
   M. Duffy and G. Waitt

Part 2: Nuts and Bolts: Making Festivals Happen

4 Local Leadership and Rural Renewal through Festival Fun:
   The Case of SnowFest 61
   A. Davies

5 Economic Benefits of Rural Festivals and Questions
   of Geographical Scale: The Rusty Gromfest Surf Carnival 74
   P. Tindall

6 Greening Rural Festivals: Ecology, Sustainability and
   Human–Nature Relations 92
   C. Gibson and C. Wong

Part 3: Politics and Place: Culture, Nature and Colonialism

7 Performing Culture as Political Strategy: The Garma Festival,
   Northeast Arnhem Land 109
   P. Phipps

8 'Our Spirit Rises from the Ashes': Mapoon Festival and
   History's Shadow 123
   L. Slater

9 Birthday Parties and Flower Shows, Musters and
   Multiculturalism: Festivals in Post-War Gympie 136
   R. Edwards

10 On Display: Ravensthorpe Wildflower Show and the
   Assembly of Place 155
   R. Mayes

Part 4: Reinventing Rurality

11 Elvis in the Country: Transforming Place in Rural Australia 175
   J. Connell and C. Gibson

12 Marketing a Sustainable Rural Utopia: The Evolution of a
   Community Festival 194
   M.W. Rofe and H.P.-M. Winchester

13 ChillOut: A Festival 'Out' in the Country 209
   G. Waitt and A. Gorman-Murray

Part 5: Festival People

14 Bring in Your Washing: Family Circuses, Festivity and Rural
   Australia 229
   A. Lemon

15 Culturing Commitment: Serious Leisure and the Folk Festival
   Experience 248
   R. Begg

16 Tartans, Kilts and Bagpipes: Cultural Identity and Community
   Creation at the Bundanoon is Brigadoon Scottish Festival 265
   B. Ruting and J. Li

17 What is Wangaratta to Jazz? The (Re)creation of Place,
   Music and Community at the Wangaratta Jazz Festival 280
   R. Curtis

Index 294

List of Figures and Tables

Figures

2.1 'Cattle parade at the Kenilworth Showgrounds', ca. 1950. Image courtesy of the Sunshine Coast Libraries 28
2.2 'BHP Exhibition at Newcastle Show', 1940. Image courtesy of the Newcastle Region Library 29
2.3 'Agricultural display at the Bundaberg show', 1912. Image courtesy of the State Library of Queensland 30
2.4 'Newcastle show: agricultural exhibit', *Newcastle Morning Herald*, 1951. Image courtesy of the Newcastle Region Library... 31
2.5 'Back to Moonta' celebrations', 1927. Image courtesy of the State Library of South Australia 33
2.6 'Shakespeare Festival', 1959. National Archives of Australia 37
3.1 Pre-concert rehearsal, Four Winds, Bermagui (photograph Gordon Waitt) 51
3.2 Drummers amongst audience, Four Winds, Bermagui (photograph Michelle Duffy) 53
8.1 Local residents getting ready for the Mapoon festival march (Photo: Lisa Slater) 124
8.2 The last remaining house that survived the burning of Mapoon (Photo: Lisa Slater) 125
9.1 Gympie's centenary procession (*Gympie Times*, 1967: 4)........ 139
9.2 Floral picture, Gympie Red Cross 'Chelsea' Flower Show 1992 (Joy Currie Private Collection)............................ 143
9.3 Advertisement for 1988 Muster (Apex Club Gympie, 1988) 148
9.4 Toyota Country: Advertisement in Muster Program, 2006 150
10.1 Show volunteers identifying wildflowers (Photo: Robyn Mayes) 163
10.2 Wildflower display, Ravensthorpe Wildflower Show, 2007 (Photo: Robyn Mayes) 165

10.3	Unsorted wildflowers, Ravensthorpe Senior Citizens' Centre (Photo: Robyn Mayes)	167
10.4	Specimens on display, Ravensthorpe Wildflower Show (Photo: Robyn Mayes)	168
11.1	Parkes, according to the *Sydney Morning Herald* (January 2003: 8) (*Source*: Brennan-Horley *et al.*, 2007: 75)	178
11.2	The CountryLink Elvis Express (*Source*: State Rail, promotional material 2005)	179
11.3	Elvis impersonators busk outside the Royal Hotel, Parkes, 2010 (Photo: John Connell)	181
11.4	'Eldest Presley', Parkes Elvis Revival Festival, 2010 (Photo: John Connell)	185
11.5	Stallholder, Parkes Elvis Presley Revival Festival 2010 (Photo: John Connell)	187
11.6	Parkes railway station, Elvis Revival Festival 2010 (Photo: John Connell)	192
12.1	Deck the Hall, Lobethal (*Source*: http://lightsoflobethal.com.au/lol/deckthehall.html 2009)	200
12.2	Solar Grove, Lobethal (*Source*: http://lightsoflobethal.com.au/lol/solargrove.html 2009)	201
12.3	Celebrating CHRISTmas as opposed to Christmas. (*Source*: Winchester, H.P.M and Rofe, M.W. 2005)	204
12.4	Welcome to Santa's Retreat ... Lock Your Car! (*Source*: Winchester, H.P.M. and Rofe, M.W. 2005)	205
13.1	Vincent Street, Daylesford (Photo: Gordon Waitt)	212
13.2	Camp 'fairy tale' float, ChillOut 2006 (Photo: Gordon Waitt)	215
13.3	Carnival Day, ChillOut 2006 (Photo: Gordon Waitt)	216
14.1	Lennon Bros Circus, 2007 (Photo: Cal MacKinnon)	230
14.2	Gills Circus/Sideshow line-up board, c. 1940s (Photo: Doyle Gill)	232
14.3	Aboriginal wire dancer, Con Colleano, c. 1930s (Photo: Topsy Hutchens)	234
14.4	Linda West, Liberty pony act. Lennon Bros Circus, 2007 (Photo: Cal MacKinnon)	235
14.5	Stardust Circus, Charleville, Queensland, 2005 (Photo: Cal MacKinnon)	237
14.6	Lorraine Ashton's Classic Circus, 2006 (Photo: Andrea Lemon)	239
14.7	Gary Grant, Lorraine Ashton's Classic Circus, 2006 (Photo: Andrea Lemon)	240

14.8 Golda Ashton (centre) with nephew, Merrick, and mother, Nikki. Backstage, Circus Joseph Ashton, 2006 (Photo: Andrea Lemon) 242
14.9 Shannan West with Cassius, Stardust Circus, 2005 (Photo: Cal MacKinnon) 244
14.10 Wonona West with Millie, Stardust Circus, 2005 (Photo: Cal MacKinnon).. 245
15.1 'Committed folkie', St Albans Folk Festival 2006 (Photo: Robbie Begg).. 256
15.2 'Night owl' jam session, St Albans Folk Festival 2006 (Photo: Robbie Begg).. 257
15.3 Blackboard Session at The Friendly Inn Pub, Kangaroo Valley Folk Festival 2006 (Photo: Robbie Begg) 258
15.4 Mediaeval dancers, St Albans Folk Festival 2006 (Photo: Robbie Begg).. 259
16.1 Caber tossing, Bundanoon is Brigadoon Scottish Festival, 2006 (Photo: Brad Ruting) 266
16.2 The parade down Erith Street, Bundanoon, 2007 (Photo: Brad Ruting).. 268
16.3 A lone piper performs in front of the tents of Clans Fraser, Macleay and Macdonald (Photo: Brad Ruting) 273
17.1 Wangaratta, 'Australia's Jazz Capital' (Photo: Rebecca Curtis). . 286

Tables

1.1 Numbers of festivals, by type, Tasmania, Victoria and NSW, 2007 6
1.2 Aims of surveyed festivals 15
1.3 Geography of inputs for festivals, Tasmania, NSW and Victoria, 2007, by geographical scale (percent of total inputs; average across all festivals) 18
5.1 Total spending by all surveyed visitors and average per visitor group in Lennox Head on the weekend of Gromfest 82
5.2 Comparison of the direct economic impacts of various Australian sports events 86

Contributors

Robbie Begg, Instinct and Reason, 20 Poplar Street, Surry Hills, NSW 2010, Australia. Email: rbegg@instinctandreason.com

John Connell, School of Geosciences, University of Sydney, NSW 2006, Australia. Email: john.connell@sydney.edu.au

Rebecca Curtis, Kellogg College, 62 Banbury Road, Oxford OX2 6PN, United Kingdom. Email: racurtis6@hotmail.com

Kate Darian-Smith, The Australian Centre, School of Historical Studies, University of Melbourne, Parkville, VIC 3010, Australia. Email: k.darian-smith@unimelb.edu.au

Amanda Davies, School of Social Sciences and Asian Languages, GPO Box U1987, Curtin University of Technology, Perth, WA 6845, Australia. Email: a.davies@curtin.edu.au

Michelle Duffy, School of Humanities, Communications & Social Sciences, Monash University, VIC 3800, Australia. Email: Michelle.Duffy@monash.edu.au

Robert Edwards, Faculty of Arts & Social Science, University of Sunshine Coast, Sippy Downs Dr, Sippy Downs, QLD 4556, Australia. Email: REdwards@usc.edu.au

Chris Gibson, Australian Centre for Cultural Environmental Research (AUSCCER), School of Earth and Environmental Sciences, University of Wollongong, NSW 2522, Australia. Email: cgibson@uow.edu.au

Andrew Gorman-Murray, Australian Centre for Cultural Environmental Research (AUSCCER), School of Earth and Environmental Sciences, University of Wollongong, NSW 2522, Australia. Email: andrewgm@uow.edu.au

Andrea Lemon, PO Box 14, Moonee Vale, VIC 3055, Australia. Email: citrus@andrealemon.com

Jen Li, School of Geosciences, The University of Sydney, NSW 2006, Australia. Email: jennigan@gmail.com

Robyn Mayes, John Curtin Institute of Public Policy, Curtin University, GPO Box U1987, Perth, WA 6845, Australia. Email: R.Mayes@curtin.edu.au

Peter Phipps, Global Studies, Social Science & Planning, RMIT University, GPO Box 2476, Melbourne, VIC 3001, Australia. Email: peter.phipps@rmit.edu.au

Matthew W. Rofe, School of Natural and Built Environments, University of South Australia, GPO Box 2471, Adelaide, SA 5001, Australia. Email: matthew.rofe@unisa.edu.au

Brad Ruting, School of Geosciences, University of Sydney, NSW 2006, Australia. Email: bradruting@gmail.com

Lisa Slater, Hawke Research Institute, University of South Australia, GPO Box 2471, Adelaide, SA 5001, Australia. Email: Lisa.Slater@unisa.edu.au

Patricia Tindall, Sinclair Knight Merz, PO Box 164, St Leonards, NSW 1590, Australia. Email: PTindall@skm.com.au

Gordon Waitt, Australian Centre for Cultural Environmental Research (AUSCCER), School of Earth and Environmental Sciences, University of Wollongong, NSW 2522, Australia. Email: gwaitt@uow.edu.au

Jim Walmsley, School of Behavioural, Cognitive and Social Sciences, University of New England, Armidale, NSW 2351, Australia. Email: dwalmsle@une.edu.au

Hilary P.-M. Winchester, University of South Australia, GPO Box 2471, Adelaide, SA 5001, Australia. Email: h.winchester@internode.on.net

Colleen Wong, Australian Centre for Cultural Environmental Research (AUSCCER), School of Earth and Environmental Sciences, University of Wollongong, NSW 2522, Australia. Email: colleen.wong@environment.nsw.gov.au

Aboriginal and Torres Strait Islander people should use caution when reading this book as it may contain images and names of deceased persons.

Preface

Festivals are a vibrant part of cultures everywhere – from traditional Aboriginal societies in Australia to London's Notting Hill Carnival; from Santa Clarita, California's annual cowboy festival (held on the film set of the cult television series, *Deadwood*) to Sydney's Gay and Lesbian Mardi Gras. With sometimes lengthy heritages, festivals have become increasingly popular and more numerous since the 1960s. Many large festivals have become cornerstones of economic development campaigns framed around tourism, especially when seeking to attract particular demographic groups with high disposable incomes; and in places as diverse as Florida in summer and Edinburgh in winter, festivals have been used to generate economic activity in otherwise quiet 'off seasons'.

While the Edinburgh festival, Glastonbury and a host of others may be world famous, most festivals are quite small and simply significant for local people and a scatter of visitors from afar. Yet, against a backdrop of rural decline, many places have sought to stage festivals to reinvigorate community and stimulate economic development. Rural festivals have consequently proliferated and diversified in recent years from the traditional country show to evermore diverse niches. How do such festivals work? Who comes, who entertains the public or runs stalls and who works behind the scenes? Do they make any money at all, or are they merely means to feel good? Are festivals really significant for rural communities despite their apparent short-lived nature? This book seeks answers to these questions.

Increasingly, festivals are seen as important to nurturing local arts scenes, new niche industries and the creative economy. They often arise from the passions of a few local people, with modest beginnings, and provide early opportunities for musicians, dancers, actors and comedians moving from amateur to professional status. This is particularly the case in smaller towns outside major metropolitan cities where performance venues may be absent and audiences tiny. In rural and regional areas of most developed countries, wider debates have emerged over the impacts of economic restructuring and the geographically uneven distribution of investment and government funding. In such debates, festivals appear to be a positive, local and organically-generated activity, with the potential to contribute to regional development by stimulating both local society and the economy.

Festivals are not without criticism. Since the original World Fairs in the 19th century, festivals have fed civic tendencies towards boosterism; have served the interests of the elite; excluded minorities and tended towards 'safe' culture; or enabled the commodification or 'packaging' of local culture for tourists naive to the potential for fabrication. More generally, as officially endorsed events, festivals always have the capacity to selectively seek and represent some elements of local cultures and identities, intensifying social exclusion – inadvertently or otherwise. In various ways, local social tensions may be refracted through festivals, as much as community is engendered.

Against this backdrop of opportunity and critique, this book explores festivals in non-metropolitan Australia. It neither suggests that festivals are panacea to all types of economic woes, nor writes them off as mere 'pastiche', commercialism or avenues for elitism. Festivals capture many of the broader contradictions and tensions that emerge when local culture becomes more fully entwined with economic and tourist promotion. Yet, as discussed in more detail throughout this book, festivals create community – in often uncontroversial ways – and are an under-acknowledged and yet potentially significant component of strategies to develop grass-roots economies. Individual festivals may not be particularly lucrative but, through their sheer ubiquity and proliferation, they diversify local economies and advance laudable goals of inclusion, community and celebration.

Irrespective of any economic benefits, rural festivals necessarily invite people to collaborate and debate what they want from their local event. In doing so, rural festivals open up opportunities for communities to wrest control of how cultural identities are constructed and economic narratives about their place are framed. This is particularly so in rural areas, typically cast as culturally 'backward' by metropolitan elites and subject to media coverage that portrays small communities as vulnerable, in decline or at the whim of metropolitan decision makers. Local involvement in staging, management and entertainment at festivals produces opportunities for counter-narratives that position local actors as central, rather than marginal, to economies of regional and national significance. Such stories of how local people have come together to stage unique and rewarding festivals are central to this book.

Preface

This book was also motivated by the perception of the limitations of much academic work on festivals. Much literature focuses on mega-events such as the Olympic Games and other hallmark global events. The multitude of small community events has largely been ignored. An urban bias also infuses festivals research, yet insights for metropolitan festivals have limited relevance in rural and regional areas. Place matters enormously when examining links between culture and economic development, and when ascertaining, for instance, the economic or environmental impacts of a festival. Rural Australia has long been maligned in the domestic media as in 'economic crisis' and facing environmental hardships such as drought and climate change. While rural and regional festivals have gained some limited attention academically, their treatment pales in comparison to that lavished on urban festivals. That may be metaphor for a wider government focus on urban versus rural centres.

Although there is plenty of anecdotal knowledge in Australia that festivals are increasingly important for rural and non-metropolitan communities, surprisingly little is known about their numbers and geographical distribution; their significance in cultural and economic terms; and whether or not they are absorbed into formal regional development and planning strategies outside major cities. While the 'science' of event economic impact assessment is now well developed, prior emphasis has overwhelmingly been on understanding the monetary balance sheet of festivals as the indicator of economic 'success' (or otherwise), or on analysing questions of cultural identity in isolation. Invariably absent has been analysis of rural festivals that brings together detailed histories and case studies, and broader integrated discussion of themes such as environment, economic impact, identity and belonging. This book seeks to achieve this kind of integration and provide a wider and more comprehensive perspective.

Naturally, thanks are due to the various contributors from many parts of urban and rural Australia, and to the many people who made it possible for us and them to undertake research, filled in questionnaires and spent time explaining the intricacies of festival management. We would love to have had more case studies from some of the quirky and extraordinary festivals that exist in regional Australia, from the Running of the Sheep in Boorowa to the Ute Muster in Deniliquin, but space is limited. We must specially thank Kelly Hendry in the Parkes Tourism Office, who has supported us since 2002 when we first visited the Parkes Elvis Festival which has now become something of an annual pilgrimage. We acknowledge the Australian Research Council who provided funding support through a Discovery Project (DP0560032), and the ARC Cultural Research

Network who sponsored a national symposium on rural festivals in which the seeds of this book were sown. We are indebted, again, to Channel View for their enthusiasm for this book and for their customary patience; to Naomi Riggs for her tireless assistance with the nuts and bolts of book editing; and to Ali Wright and Cara Gibson for forbearance and support at home.

Chris Gibson and John Connell
June 2010

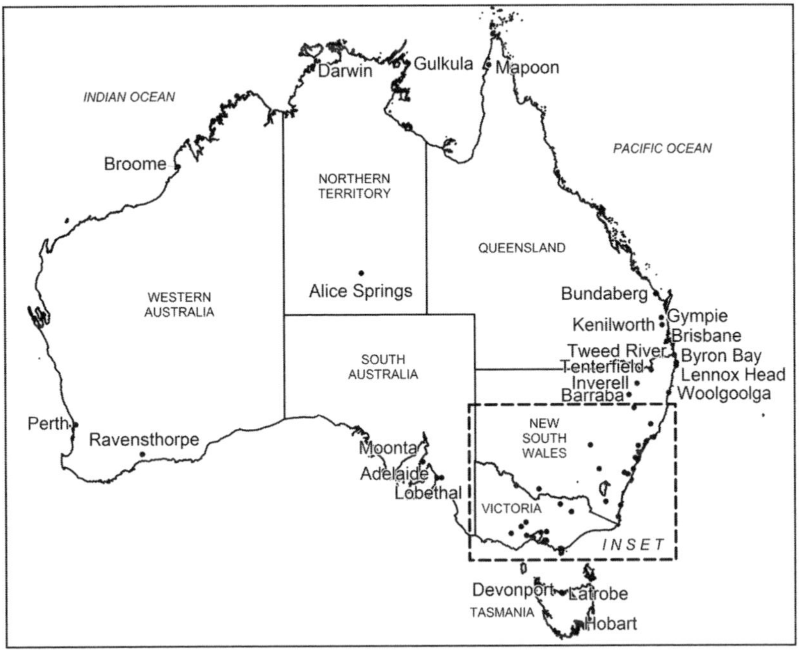
INDIAN OCEAN
PACIFIC OCEAN
Darwin
Gulkula
Mapoon
Broome
NORTHERN TERRITORY
QUEENSLAND
WESTERN AUSTRALIA
Alice Springs
Bundaberg
Gympie
Kenilworth
Brisbane
Tweed River
Byron Bay
Tenterfield
Lennox Head
Inverell
Barraba
Woolgoolga
SOUTH AUSTRALIA
Perth
Ravensthorpe
Moonta
Adelaide
Lobethal
NEW SOUTH WALES
VICTORIA
INSET
Devonport
Latrobe
TASMANIA
Hobart

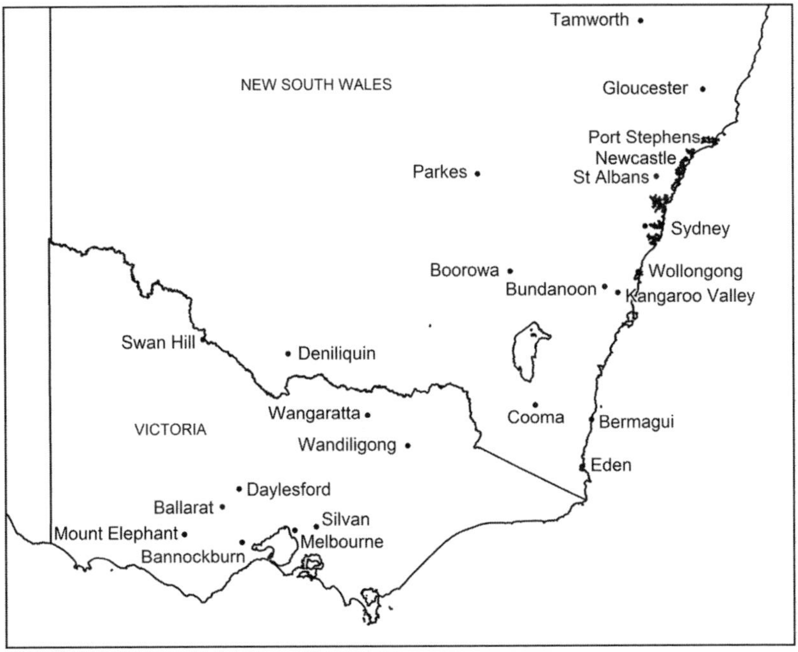
Tamworth
NEW SOUTH WALES
Gloucester
Port Stephens
Newcastle
Parkes
St Albans
Sydney
Boorowa
Wollongong
Bundanoon
Kangaroo Valley
Swan Hill
Deniliquin
Wangaratta
Cooma
Bermagui
VICTORIA
Wandiligong
Eden
Daylesford
Ballarat
Silvan
Mount Elephant
Melbourne
Bannockburn

Part 1
Exploring Rural Festivals

Chapter 1
The Extent and Significance of Rural Festivals

C. GIBSON, J. CONNELL, G. WAITT and J. WALMSLEY

Festivals are enjoyable, special and exceptional, sometimes the only time of celebration in small towns. Festivals are full of rituals of entertainment, spectacle and remembrance, and they bring people together. Most people participate for enjoyment, something different and the pleasure of coming together. Festivals offer much more than just dollars and cents, or place-marketing and branding, although both are implicated. Festivals create culture, engaging some but excluding others. The narratives of participants may articulate a strong sense of being part of a community, however transient, or a reframing of a personal understanding of a specific issue, perhaps sustainability, multiculturalism or reconciliation. More often festivals simply enhance already existing pleasures, from beer drinking to line dancing. Festivals take multiple forms and play multiple roles.

This chapter summarises the findings from a three-year Australian Research Council (ARC) funded project, which sought to document the extent and significance of festivals for rural communities and economies. Hitherto, no studies have sought to measure the geographical extent of festivals, despite numerous case studies done usually from a single disciplinary perspective. What was missing was a sense of the diversity of economic and socio-cultural impacts of festivals in rural localities – understood cumulatively as well as individually. Hence, we sought to develop a different approach, based on accumulating information about festivals across a large area, and building a stock of basic information on their operation and how they nestled within rural communities. We were interested in a range of questions: the role of festivals in rural and regional restructuring; on post-agricultural or 'post-productivist' transitions in rural places; and the changing nature of rural cultural identities. Eventually, we constructed the largest ever database of rural festivals in Australia, with more than 2850 festivals in just three eastern Australian states – Tasmania, Victoria and New South Wales – and through wide-ranging qualitative research with a small, selected number of rural festivals, as well as completed questionnaires from 480 festival organisers (Gibson et al., 2010), gleaned insights on the ability of festivals to catalyse social and community networks, to inform regional development and to challenge or sustain rural cultural identities.

Festivals were located mainly by using internet search engines, and running detailed queries by keyword for every local government area (LGA) in the three states, and for particular niches (e.g. for particular styles of music), for common festival types (e.g. food and wine festivals) and for more specific activities associated with demographic groups, subcultures and other leisure activities (e.g. 'hot rod' car shows, seniors festivals, gay and lesbian festivals, goth festivals, and diverse sports). Festivals were also located via print media (including regular scanning of metropolitan broadsheets), regional tourism brochures and flyers. Under-represented were festivals without a formal organising committee, postal or email address, and festivals such as those organised by neo-Pagans, Radical Faeries or similar-minded 'New Age Travellers' who were seeking to operate in social networks outside mainstream society (see Begg; Gibson & Wong; Slater, this volume). The database and survey eventually became a baseline for various studies that have become chapters in this book. Subsequent detailed studies included analysis of economic dimensions via visitor and business surveys at the Elvis Revival Festival in Parkes, in inland NSW (see Chapter 11), at ChillOut, Australia's largest rural gay and lesbian festival, held in Daylesford, Victoria (see Chapter 13) and at Gromfest, a youth surfing festival (see Chapter 5). Additional case studies were pursued elsewhere based on related themes, including cultural identity (see Chapter 16), and environmental sustainability (see Chapter 6). In this chapter, we review the overall results of the project and in doing so introduce key themes and issues that resonate throughout the remainder of this book, and that have drawn in contributors with quite different perceptions of the role of festivals in diverse contexts.

Much debate surrounds what constitutes a 'festival'. A demarcation was made between infrequent, usually annual events, which were included (pending other criteria), and regular, recurrent events (such as sporting fixtures and regular musical nights) held throughout the year that were excluded. Conferences, conventions and trade exhibitions were also excluded. Generally festivals had to meet at least one (and preferably more than one) of the following criteria: use of the word 'festival' in the event name; being an irregular, one-off, annual or biannual event; emphasis on celebrating, promoting or exploring some aspect of local culture, or being an unusual point of convergence for people with a given cultural activity, or of a specific subcultural identification. Festivals then were understood following Getz (2007: 31) as 'themed, public celebrations'. Occasional festivals were spectacular, evoking an older notion of carnival, where 'people do something they normally do not; they abstain from something they normally do; they carry to the extreme behaviours that are usually regulated by measure; they invert patterns of daily social life' (Falassi, 1987: 3). This was rare, but evident at the Daylesford ChillOut Festival (see Chapter 13). At most rural festivals people focused on very specific activities (as enthusiasts of a certain sport, music style, collector's habit or pastime such as gardening), and simply did more of what they did at other times, sometimes flamboyantly – eating, drinking, dancing, singing, spending. What makes festivals different from other events is that they were usually held annually and generally have social aims: getting people together for fun, entertainment and a shared sense of camaraderie. Most festivals created a time and space of celebration, a site of convergence outside of everyday routines, experiences and meanings – ephemeral communities in place and time.

The Diversity of Rural Festivals

Festivals are increasingly numerous, and kaleidoscopically diverse; rural and regional Australia has never hosted as many festivals, an indication of evolving creativity and ingenuity (Gibson *et al.*, 2009; Gibson, 2010). This diversity of festivals reveals the cultural diversity of non-metropolitan Australia – a cultural diversity not so much about ethnic polyphony in the conventional sense (as measured by international migration) but about diversity of cultural pursuits and ideas. The most common festivals were sporting, community, agricultural and music festivals – which together made up three-quarters of the festivals (Table 1.1). Within these categories there was further diversity. 'Community' festivals covered everything from Grafton's historic Jacaranda Festival (named after the town's signature tree) to Kurrajong's Scarecrow Festival, Nimbin's Mardi Grass (a marijuana pro-legalisation festival), Ballarat's Stuffest Youth Festival, Ettalong's Psychic Festival, Tumut's elegant Festival of the Falling Leaf, Queanbeyan's Festival of Ability, and Myrtleford's Tobacco, Hops and Timber Festival. Similarly varied were sports festivals, covering

Table 1.1 Numbers of festivals, by type, Tasmania, Victoria and NSW, 2007

Type of festival	TAS	VIC	NSW	TOTAL**	% of total
Sport	86	485	488	1059	36.5
Community	45	216	175	436	15.0
Agriculture	19	146	215	380	13.1
Music	13	116	159	288	9.9
Arts	12	73	82	167	5.8
Other*	7	87	71	165	5.7
Food	10	53	67	130	4.5
Wine	7	49	32	88	3.0
Gardening	20	43	14	77	2.7
Culture	2	21	11	34	1.2
Environment	1	8	12	21	0.7
Heritage/historic	4	8	7	19	0.7
Children/Youth	0	10	5	15	0.5
Christmas/New Year	0	10	2	12	0.4
Total	226	1325	1340	2891	100

Source: ARC Festivals database, 2007
*The other category includes small numbers of the following festival types: Lifestyle, Outdoor, Science, Religious, Seniors, Innovation, Education, Animals and Pets, Beer, Cars, Collectables, Craft, Air Shows, Dance, Theatre, Gay and Lesbian, Indigenous and New Age.
**The total for this table is slightly more than the total number of festivals in the database, due to counting of some festivals in more than one category. This occurred when separating categories proved impossible (e.g. for 'food and wine festivals').

everything from fishing to billy carts, cycling, pigeon-racing, hang gliding, track and field events, horse racing, basketball carnivals, ski races, dragon boat racing and campdrafting.

Places with the most festivals tended to be large regional towns outside capital cities (such as Newcastle, Wollongong and Wagga Wagga), regions reliant upon tourism industries (Snowy River and Coffs Harbour) or coastal 'lifestyle' regions with mixes of tourism and retiree in-migration (Lake Macquarie, Bass Coast, Great Lakes). Such places have large enough populations to justify a sequence of festivals with minimal risk of failure.

Moreover, several are within 'day tripper' driving distance of Sydney and Melbourne, and most have dedicated regional tourism offices, major tourist information centres and expertise in place marketing (Gibson et al., 2010). Some 'festival capitals' emerge – like Ballarat, with 73 festivals across the calendar year – where there is a clearly identifiable, professional festival and event management industry, with rare parallels in smaller towns (such as Daylesford and Byron Bay).

Few festivals were organised as subversive protests, despite the rapid social and economic transformations beyond metropolitan centres. Exceptions included the counter-spaces of folk festivals (Chapter 15) and 'free-festivals', such as the Homeland Festival of Sacred Song and Dance (Dorrigo) and Woodford Festival (where music combines with poetry, politics, Koori (Aboriginal) ceremonies, open-mike sessions and workshops on sustainability and activism). While satire may be an integral part of many street parades – such as the Daylesford New Year Festival (which typically, and unusually for a regional festival, features floats that comment on local controversies and parody big corporations) – rarely were festivals opposed to political systems, or offered different political agendas or to win new loyalties. Nor did festivals resonate with Bakhtin's (1984) concept of the 'carnivalesque', where festivals are sanctioned by the political elite as a mechanism to help the masses forget social injustices and inequalities: bread and circuses. Where rural communities have engaged in political action, on issues as diverse as the closure of regional services and infrastructure, timber cutting, climate change and industry deregulation, they have usually resorted to more formal protests, petitions and lobbying (Gibson et al., 2008). Instead, where politics featured at rural festivals in our research, this was usually a somewhat reserved integration of political issues with more simple pleasures: a stall from the local branch of the Green Party or the National Farmer's Federation amongst those selling bags, t-shirts, hot-dogs and beer, or charities promoting 'good causes' such as ecology, sustainability or breast cancer research as part of a festival's aims and activities (see Chapters 6 and 10).

Country, jazz, folk and blues festivals counted for over half the music festivals – far outweighing styles such as rock or hip-hop that are more commercial or lucrative in the wider retail market for recorded music (Gibson, 2007). Music and arts festivals were more vernacular than elite: sewing and quilting festivals were as common as opera festivals; country music was more prevalent than jazz, and 'roots', bluegrass and folk were more popular than commercial sales of recorded music might suggest. Music festivals were amongst the largest of the festivals, including Tamworth's annual country music festival, Tweed Heads' 'Wintersun'

Rock and Roll/1950s nostalgia festival, and Byron Bay's East Coast Blues and Roots Festival and Splendour in the Grass, where audiences ranged from 15,000 to 100,000 people.

The vast bulk of regional festivals were otherwise comparatively small. The average attendance at festivals was 7000, but the actual sizes varied enormously. The Victorian Seniors Festival, actually held in many locations at different times throughout the state, claimed an attendance of 400,000. By contrast, the tiny Summit to the Sea endurance cycling festival had a mere 15 participants. Some 138 festivals (29%) had audiences of fewer than 1,000 people and two-thirds had fewer than 5000, while just 11 festivals (barely 2%) had audiences of more than 50,000; four of these were agricultural shows and another four were music festivals.

Generally small numbers indicate both that country towns themselves are small and that many festivals reflect the proliferation of specialist niches for people with shared passions – whether pigeon-racing or model train collecting. At such events, subcultural or even 'neo-tribal' relationships are formed between people with shared enthusiasms who speak the same (often arcane) language (Chapter 15). More frequently small numbers reflected small places and thinly spread populations – demographic reality making it difficult to generate large crowds. In these small festival settings, the range of sometimes socially diverse people who live nearby are brought together into what are often quite intimate social settings. Anonymity is near impossible, with participants personally encountering entertainers and activities, friends and neighbours, and even sworn enemies. More than anything else this sense of totally embracing community distinguishes rural and urban festivals.

Festivals thus provide unique opportunities to bring unrelated people together in novel ways (see Chapter 3), but also reinforce pre-existing relationships and social ties. Because rural festivals are on the whole small and geared towards local communities, rather than tourists or other outsiders, they provide venues for individuals, families and friendship-circles to sustain and re-invigorate emotional ties. In the unusual case of the ChillOut Festival in Daylesford, the event built an alternative understanding of community by providing a context for local residents to challenge their assumptions about sexuality, and dispel the anxieties of some local people over homosexuality in rural Australia, however tentatively (see Chapter 13). Few alternative instances exist and are as successful as community festivals in bringing together, however fleetingly, otherwise disparate social groups. Through their capacity to evoke emotions, festivals thus have the capacity to reconsider and re-imagine what constitutes the local community.

While newcomers like the ChillOut Festival have been in place for just 14 years, 67 festivals who returned a survey had been running since before 1900. These were mostly gardening and flower festivals (a mainstay of Australian country life), and rodeos agricultural shows, an archetypal festival in Australian country towns, traditionally focused around harvest, produce, livestock and new farming technologies. In the post-war period agricultural shows became much larger, featuring live music (often country music), fairground rides, showbags (bags of toys and sweets usually bought by or for young children) and commercial stalls (see Chapter 2). Many agricultural shows however lost their impetus, closed down or amalgamated with others nearby; those that had survived were either very large or found other ways to remain relevant, including niche marketing and hiring big name music acts. Future challenges are presented by ageing committees and succession and, of all rural festivals, this was the only group experiencing decline.

Unlike agricultural shows, most festivals had their genesis in the past 30 years. A third of the festivals began after 2000 and 61% began after 1980. This festival flurry was characterised by a much greater diversity and novelty in festival themes. As many as 88% of the music festivals, 84% of the food festivals and 87% of the wine festivals began after 1980: an indication of both agricultural transformation in regional Australia, and the rise of new forms of consumption and recreation. Adversity in rural Australia has been brought about through rapid social and economic change, because of a mix of neoliberal politics, economic globalisation, communication technologies, drought, retiree in-migration and new 'sea-change' and 'tree-change' migration from metropolitan centres, that brought a new rural population structure and new economic activities (Burnley & Murphy, 2004; Connell & McManus, 2010). At festivals, opportunities exist to rearticulate identities amidst this sense of adversity: be it a particular expression of country-mindedness, or reviving what it means to be living beyond the metropolitan centres. For some (particularly baby boomers in early retirement), leisure time has increased, and regional Australia has become more accessible to metropolitan centres, coinciding with rural areas diversifying to become 'post-productivist' countryside. Festivals consequently also became more prolific, and diverse.

Putting Places on the Map

While larger regional centres host the most festivals, some quite small towns also sponsored many festivals, especially in the Riverina region (surrounding the Murray River in NSW and Victoria) and around the

Snowy Mountains. Such patterns were probably a function of the willingness of local residents to get out and support festivals in their home town; central support from councils that employed festival and event officers; and regional 'contagion' effects (where neighbouring towns gain inspiration or steal ideas from each other). In some cases, festivals have been a response to change, a strategy to mitigate economic crisis in agriculture, and an outcome of civic desires to 'do something' to reverse the sense of decline in their towns. In many of the most festival-dense towns in the Riverina region, agriculture has all but collapsed following rationing of water for irrigation (in a region heavily dependent upon it for industries such as citrus and rice) and closures of schools and railways and rationalisations of government employment have exacerbated already-present trends of shrinking population and ageing. Festivals can be an overt means to reverse the demographic and psychological damage of decline. Some small towns, actively seeking to boost their populations, deliberately use festivals as a means of advertising themselves to metropolitan populations, by promoting more congenial lifestyles (Connell & McManus, 2011). Tenterfield, in northern NSW, has deliberately sought to develop a wine industry aimed at metropolitan consumers, as an export market, but also as a place-branding strategy. Its annual wine festival is critical because, in the words of its organiser, it 'brings as many people as we can from outside Tenterfield,' and we say 'this is what Tenterfield is' (Eaton, 2007) – a cosmopolitan, culturally rich place. In-migration of middle-class 'tree-changers' from capital cities has indeed ensued.

Barraba in NSW was once a vibrant asbestos-mining town, but for obvious reasons fell into crisis with mine closures. Rather bravely this town of less than 1200 people has since sought to reinvent itself as place for bird-watching (via dedicated 'bird routes', transformed from old droving livestock routes), fossicking and festivals. The town has also flirted with themed festivals for jazz and country music (respectively billed as 'Australia's smallest jazz music festival' and 'Australia's smallest country music festival') and, for antiques enthusiasts, even a festival dedicated to fine chinaware. Festivals might not alone be able to reverse the outcomes of macro-economic restructuring, but make it possible to present images of vibrancy and community to an otherwise apathetic outside world.

Festivals can place or keep towns on the map, and are a great way to market places, often more effectively than bland official branding strategies. Marketing through festivals enables celebration of nature, local produce and industry, seasonal transitions or other cultural traits and creates images that may linger in the national imagination: the Collector Pumpkin Festival; Lithgow's IronFest; Guyra's Lamb and Potato Festival;

Gundagai's Dog on the Tuckerbox Festival. Other places have become 'festival places' less because of any link to local uniqueness, but because of famous, iconic or 'authentic' festivals held there that have become legendary within particular national (and international) audiences of specialist fans: Port Fairy (and its folk festival), the Wangaratta Jazz Festival (see Chapter 17), the Tamworth Country Music Festival (see Gibson & Connell, forthcoming) and the Warwick Rodeo (Hicks, 2003).

Some towns have been able to gain significant economic and social benefits by developing and trading on unlikely, improbable, even wholly fictitious and sometimes 'unworthy' events and associations. At Tamworth, an association with country music (that most iconic of American music styles) was carefully cultivated in the 1980s and has since become pervasive. At Bundanoon, a mere similarity in pronunciation with the iconic Scottish Brigadoon led to a Scottish festival which has over time become renowned as an 'authentic' celebration of heritage and ancestry (see Chapter 16). In South Australia, the residents of Lobethal celebrate a more legitimate history (see Chapter 12). The very temperate town of Gloucester annually ships in snow from several hundred kilometres away for its improbable summer SnowFest (see Chapter 4). At Deniliquin, its annual Ute Muster has put it on the map so much so that the town is considering investing in a museum dedicated to the iconic Australian utility vehicle. At Boorowa's Irish Woolfest, the somewhat wacky, yet brilliant idea to replicate Pamplona's 'running of the bulls' as a 'running of the sheep' down the town's streets, has earned the town national television coverage and greater patronage. Other festivals have focused on scarecrows (Kurrajong), beards (Glen Innes), chocolate (Latrobe), tomatoes (at Gunnedah's National Tomato Contest), mud (Bulli) and metal (the music style – at Scone's Metal Stock festival). Creativity has excluded few subjects.

Frequently festivals link into attempts by towns to claim national (and international) prominence as 'capitals': Australia is Tamworth's 'country music capital'(Gibson & Davidson, 2004); Deniliquin now calls itself the 'ute capital of the world'; Coonabarabran is the 'astronomy capital of Australia' (hosting an annual Festival of the Stars); Canowindra, with a population of less than 2000, is the 'ballooning capital of Australia' (hosting one of Australia's largest ballooning festivals); Gunnedah is Australia's self-proclaimed 'koala capital'; Daylesford, Victoria is Australia's gay capital (see Chapter 13) and Alice Springs, in remote central Australia is 'Australia's lesbian capital' (and hosts the Alice In Wonderland Gay and Lesbian Fun Festival) while both Casino (NSW) and Rockhampton (Queensland) are caught in a contest for the right to call themselves the 'beef capital of Australia', with both hosting annual Beef Week festivals.

When Albion Park, in Wollongong (NSW) topped an online poll of Australia's most 'bogan' places ('bogan' being a derisive term depicting people of working-class origin as 'uncultured', much like 'white trash' in the United States), the town responded by proposing that it host an annual bogan festival, as 'Australia's bogan capital'. The Parkes Elvis Festival (in 'the Elvis capital of Australia') demonstrates how a small, relatively remote place can stage a festival that generates substantial economic benefits, fosters a sense of community, without any conceivable local claim to musical heritage (see Chapter 11). Yet the festival has invigorated the town, and attracted increasing numbers of visitors on a hot weekend in the tourist off-season, because it is well-organised, slightly weird, in a friendly town and above all, fun.

Revising Rural Place Identities

Festivals can act as mirrors, reflecting a particular collective identity of a place through their program of events and activities. On the one hand, there may always be residents who do not share the love, passions, activities, politics and entertainment of the festival organisers and participants. Some become vocal opponents citing 'noise', 'crowds', the disruption of daily life and the challenge to 'proper' notions of local identity (at was the case early in the histories of both the Tamworth Country Music Festival and the Parkes Elvis Revival Festival). On the other hand, many residents wait with eager anticipation for festivals, and work to promote them, while residents and visitors, brought together by the event, may leave with a stronger sense of collective identity forged through participation. Festivals operate simultaneously as places for pleasure and the performance of identity. Enjoyment of a particular music, sport or handicraft can operate as an emotive link between people from otherwise different backgrounds. In contrast festivals may provide a moment to reflect on what it means to be 'rural', and to reinforce or revise these meanings accordingly. At some festivals – such as Garma in the Northern Territory – the activities concern critical social issues (reconciliation, political interventions in the running of remote Aboriginal communities), and sometimes overtly challenge assumptions and myths of rural place identity (see Chapter 7). At Mapoon, in Far North Queensland, a tiny Aboriginal festival had nothing to do with tourism or place promotion, but was conceived as an opportunity for the community to demonstrate pride in its history and survival in the face of oppression; that for at least one day of the year it was not a far-flung and forgotten relic of colonial past, but a living place, the centre of the universe for a small number of people (see Chapter 8).

At agricultural and flower shows 'settler' societies conceive of themselves and their colonial pasts in the present (see Chapters 2, 9 and 10). At the Australian Celtic Festival and Scottish Highland Gatherings traditions and pasts are invented (Connell & Rugendyke, 2010; Chapter 16). Organising committees take responsibility for retelling and retailing the past, and festivals help cement and/or challenge what is means to be an Anglo-Celtic or European Australian in a particular country town. Such festivals are organised not solely because of the ability to toss a caber, or consume schnapps, but because they enable a particular expression of 'settler' Australian identity in a post-colonial society – as 'country', 'home', 'the rural', 'outback' or 'bush' – even if largely invented, or embellished with nostalgia (Gibson & Davidson, 2004; see Chapter 12). Pleasure is derived from using retrospective and nostalgic perspectives to forge a temporary haven from the world.

In contrast to festivals that cherish indigenous or migrant settler identity (or both, in the case of Broome's Shinju Matsuri Festival), the Deniliquin Ute Muster relies upon celebrating an Australian icon, the utility vehicle. Pleasure and pride centre on ideas about Australian rurality as masculine and Australian nature as tamed by men and their machines (the 'ute' being the primary vehicle of working farmers). The festival strengthens allegiance to a national pride through celebrating the values of mateship, larrikinism and competition. Events include setting world records for the number of people wearing a Chesty-Bond blue singlet (an iconic piece of Australian workwear), the number of utes in a row, as well as a host of competitions including a 'Grunt-Off' (teams pulling a ute as fast as possible). The organisers warn, or indeed celebrate, on their web-page that 'political correctness has not made it across the Dividing Range', to a festival able to boast being the single largest point of sale of the nation's leading brand of rum. Beyond the souped-up engines, alcohol and testosterone, is another, more complex story about what constitutes rural Australia: while seeking to embody what it means to be 'rural' in Australia, the Deniliquin Ute Muster is very *atypical*; it is both modern (in vehicles and beer), blue-collar (no gourmet food or wine) and celebratory of a kind of iconic rural life that (despite the 'modernity' on show in customised cars) is fading into the past.

Local Heroes

Most festivals were quite local in orientation, and relatively few sought to emulate the success stories of Parkes, Deniliquin or Tamworth, in becoming tourism ventures with audiences from far beyond the local community.

The story of Australian rural festivals is of many thousands of quite tiny festivals serving their local communities as points of celebration, community and identity, in a much more prosaic manner, drawing on local expertise and volunteers, and supporting and being supported by small local businesses. Every weekend across Australia literally hundreds of small festivals take place in community halls, showgrounds, public parks and pubs. Most are modest, advertise locally, are unknown anywhere else and seemingly insignificant. But cumulatively, festivals are a vibrant, ever-present grass-roots community sector.

Collectively across the festivals we surveyed, 59% of participants were from the immediate locality and 11% were from the state capitals (Sydney, Melbourne or Hobart). Most of the rest (21%) were from elsewhere in the state, a handful came from interstate (8%) and a tiny 1% were international visitors. Just as participants were mainly local, so were the organisers. The vast majority (74%) of festivals were run by local non-profit organisations, usually tiny in size, and only 3% were run by private sector/profit-seeking companies. Reflecting this, the stated aims of festivals were often linked to the pastimes, passions or pursuits of individuals on organising committees (hence, one Tasmanian festival's key aim was 'To bring awareness of the heritage and culture of wooden boats and celebrate the skills and traditions'), or to socially or culturally orientated ends such as building community (75 festivals, or 16%), rather than as income-generating ventures (Table 1.2). Indeed, of all categories of festival aims, 'to make money' or 'to increase regional income' were the rarest responses (recorded in only 5% of cases). Promoting or showcasing a region or locality, or showcasing local products (exclusive of agricultural shows) accounted for 46% of all the festivals. Related goals such as celebrating local culture, heritage, cultural diversity, the environment, Aboriginal culture, sustainability and health were recorded by 60 festivals (or just over 10% of the stated aims). Even sporting festivals, whose objectives involved 'competing', were often quite small and also sought to 'foster and encourage talent'. So did many music festivals, such as Hobart's Festival of Voices, the nation's largest choir festival, described by the organisers simply as 'inclusive, non-competitive and affordable'.

Nearly three-quarters of all festivals were supported in some way by their local council, usually through advertising (62%), mainly on their events calendars and websites. Some 38% of the festivals with council support received financial contributions, largely as small grants. Less frequent forms of council support included providing staff support, free or subsidised venues, assistance with waste management and coordination

Table 1.2 Aims of surveyed festivals

Aim	No. of festivals*	% of festivals surveyed
To promote a place/theme/activity	137	28.5
To show(case) a place/theme/activity	86	17.9
To build community	75	15.6
To compete	75	15.6
To entertain	65	13.5
To foster/encourage	63	13.1
To celebrate	44	9.2
To fundraise	41	8.5
To educate	21	4.4
To make money	12	2.5
To increase regional income	12	2.5

Source: ARC Festivals Project survey, 2007
*The total number of festivals by aim is greater than the total number of surveys received. Festivals that recorded more than one aim are counted for each of these records.

of volunteers. Despite so many receiving some council support, only 24% of all festivals indicated that they were part of long-term economic development strategies. Rarely are festivals positioned by councils in economic terms, components of regional economic development strategies or even part of tourism development strategies.

Some festivals help to redefine the nature of life in the places where they occur. Others build on the life that presently exists. On balance, a larger percentage of festival organisers believed their festival offered something new (41%) rather than building on the existing way of life (25%). Only 7% of respondents believed their festival both offered something new and built on the existing way of life. Types of festivals that tended to indicate 'offers something new' were music, community, food and film, perhaps all part of the emerging post-productivist countryside, rather than the agricultural shows and sporting events. For most organisers their festivals had local roots. Some 60% believed their festival was reliant on the local environment either 'quite a lot' or to a 'large extent', while only 6% believed their festival was not at all reliant on the local environment. Festivals in rural and regional Australia are deeply connected to geography: they are expressions of local places, as well as local people.

With few exceptions, the small scale and local objectives of rural festivals pose challenges to the idea that festivals are seeking to re-profile the place-image of a town or region (see Getz, 1991). Rural festivals are instead mostly forged by small committees of enthusiasts who are simply seeking to have fun and commemorate their town amongst themselves and with others.

Dollars and Cents

Beyond their social and cultural motivations, festivals are also lively cells of economic activity, particularly so in small places where their relative monetary impact is greater than in urban areas (see Chapter 5). While few rural festivals make significant sums of money, seemingly small and insignificant festivals (when judged in monetary terms) can cumulatively generate notable amounts of income and employment, and catalyse qualitative benefits for a local economy. For logistical reasons it was impossible to undertake a detailed input–output or cost–benefit analysis of every festival in our database or survey, which in itself is a challenging exercise. However, festival organisers maintain their own estimates of profit or loss, turnover and funding support.

Rural festivals had small funding bases, limited turnovers and frequently only just broke even or made very modest profits. This reflects the aims of festival organisers which were rarely about making money or increasing regional income. Some 63% of festivals had a turnover in the minimum range of $0–50,000, while an even greater percentage (82%) of these small festivals received funding support. Only five rural festivals generated more than $5 million in direct turnover. The largest, most successful festivals deliver other benefits to communities, such as new facilities that can be used year-round. Tamworth is the ultimate example: largely on the basis of its iconic country music festival, Tamworth has received investment in tourism infrastructure, new hotels, museums, a major tourism information centre, and a large regional entertainment centre (securing NSW state government funding) (see Gibson & Connell, forthcoming). In Parkes, its growing Elvis reputation brought a new permanent museum (Chapter 11).

It is unsurprising that over three-quarters of festivals (78.4%) operated at a loss, broke even or made less than $10,000 in profit. But festivals can generate economic benefits for their local communities. In 2009, with an estimated attendance of 5,500, the Parkes Elvis Revival Festival brought in $3.5 million in direct visitor expenditure (at an average of $643 per visitor). In 2008, with an attendance of 22,000, the Deniliquin Ute Muster generated direct visitor expenditure of $13 million (at an average of $610 per person). Even in much smaller festivals the impact is notable: at the Gromfest youth surf carnival in Lennox Head, Northern NSW, 1200 visitors attended, spending $472 per person on average. This translated into nearly $600,000 of spending, injected into the small town over the course of a weekend (Chapter 5). Using data generated by economic modelling of visitor expenditure at five festivals, mainly larger ones, it was possible to estimate their total economic activity, and extrapolate from that. More than $500 million was generated by festivals in turnover (through ticket and merchandise sales), while from mean visitor spending figures and attendance data, across the 2856 festivals as much as $10 billion may be generated for local economies.

Despite few individual festivals making much money, they catalyse monetary benefits for their surrounding communities as a flow-on effect, through tourism visitation expenditure, the hiring of local expertise, services and materials. Benefits are felt most by an array of small businesses functionally connected to the festival, such as cafes and restaurants, sound and lighting equipment hire, waste management, hotels and motels, pubs, printers, advertising agencies, legal services and catering companies. In other words, festivals act as a local 'matchmaking' service, connecting and reconnecting local businesses. Geographical links between festivals and places are particularly evident in such local inputs (Table 1.3). For certain activities, such as staffing, catering and staging, reliance on the local economy was almost total; few of these inputs came from outside the local economy so that the multiplier effects of the festivals were quite high. The story was somewhat different for stallholders and 'talent' (musicians, performers, contestants, etc.): although the nearby areas were still the most common source of these inputs (64% and 57% respectively), they were somewhat 'less local' than for other inputs. Indeed a well-established network of itinerant stallholders travel from festival to festival in seasonal circuits, earning a living selling food, clothing or other items (something of an echo of earlier travelling shows, circuses, rodeos, boxing tents and side-shows: see Chapters 2 and 14). Some festivals had consequently made decisions to deliberately exclude such stallholders, who are frequently stigmatised as nomadic 'outsiders', to protect local businesses, schools and charities (who run the cake stalls, barbeques and merchandise sales) or to exert greater control over what food and merchandise items were offered for sale. Trying to 'keeping money local' is a priority, whereas bringing at least some talent (headline bands or foreign

Table 1.3 Geography of inputs for festivals, Tasmania, NSW and Victoria, 2007, by geographical scale (percent of total inputs; average across all festivals)

	Local (<50 kms)	Capital city in that state	Elsewhere in the same state	Interstate	International
Staff (for organizing and running the festival)	90.6	4.4	3.2	1.5	0.1
Catering (food and drink)	90.6	2.4	6.6	1.1	0
Staging (e.g. PA systems, tents, seating)	84.3	5.7	7.6	2.3	0
Stallholders (e.g. souvenirs, merchandise)	64.0	6.1	23.9	6.0	0.3
Talent (e.g. bands, competitors, performers, artists)	56.6	11.6	20.2	9.6	2.0

Source: ARC Festivals Project survey

films) from a distance is a valuable marketing tool. While economic gains may ultimately be small, festivals have enabled renewed investment in such things as public parks, car parking areas, tourism information centres, and music and performance venues, so that there are broad community benefits.

Hives of Activity: Employment Creation

While there are challenging methodological questions attached to estimating the employment generation effects of festivals (Gibson *et al.*, 2010) it was readily evident that festivals were valuable sources of new jobs in places often suffering from drought, neglect or economic decline. From detailed results from 480 festivals, and extrapolating this for all 2856 festivals, it was estimated that as many as 176,560 jobs are created directly in the planning and operation of rural festivals. On average 4.1 full-time jobs were directly created in each festival in the planning stage, alongside 5.1 part-time jobs, while 13 full-time jobs and 12.6 part-time jobs were created over the duration of the festival. Across all festivals, some 99,448 jobs were directly created in planning and running the festival. The most common were event managers/directors/coordinators (25% of jobs created), administration and accounting positions (24%), ground-keepers, ground staff and facilities managers (12%), public relations, promotions and marketing positions (9%) and artistic services (including artists, artistic and musical directors – 8.5%). Other paid positions created by festivals included retail staff, cleaners, security, catering, judging, stage crew, announcers, and tourism and community development planners. Festivals thus appear to produce around 40,000 more jobs in regional New South Wales, Victoria and Tasmania than agriculture, once the mainstay of each of these states.

Indeed, festivals are deceptively effective creators of local jobs. While individual festivals generate intermittent jobs at the time of a festival, almost all supported some full-time, year-round planning work; while where larger towns had several festivals, such as at Ballarat, festivals sustained year-round employment across a diverse range of support services. Larger festivals such as those at Tamworth and Parkes had their own dedicated staff. Moreover for some labour categories that might seem contingent only on the running of individual festivals – security, sideshow attendants, catering, judging, stage crew, announcers – the actual employment was more stable and year-round, because many of those working in such jobs travel from town-to-town in a circuit of related festivals.

Beyond paid employment there is massive volunteer support, an important reflection of their community-building role. Festival organisers estimated that 19.2 days were spent by the average volunteer assisting their festival during its planning phase, and 5.7 days on average assisting during the running of the event at time of operation. Across 480 festivals this constituted the equivalent of over 8600 days (or 23 years' worth of labour) when adding up the work done by the average volunteer across all festivals in a calendar year, which then suggests more than 70,000 days' worth of volunteer labour across the 2856 festivals in our wider database. If even modest estimates are made about the numbers of volunteers working on each festival, the resulting numbers on the extent of volunteerism are immense: if the average festival was assumed to have 20 volunteers (and many festivals had a lot more), the input was more like 1,422,000 days (or 3900 years) of labour. Volunteers are thus critical for the smooth running of festivals. Most take part in 'on the day activities' including security, catering, judging, stage crew and marshalling. Others help to set up the festival, and clean up afterwards. Without them many festivals would simply fade away so that retaining their support and loyalty is absolutely crucial. In this way festivals are thoroughly embedded in local society.

Together in Hard Times

At the time of this work, much of south-eastern Australia had experienced the longest and most severe drought in living memory. Indeed 70% of the festival organisers indicated that their community had been affected by drought, particularly in its impact on agriculture (32%) resulting in economic downturn for communities reliant on the agricultural sector (31%). Some 17% noted that their community had been impacted by water restrictions, affecting many aspects of community life. The impacts of the drought were environmental, economic and social, and especially in inland areas had reduced community morale and increased economic stress.

Almost half the festivals (43%) were affected by drought. Of those the most common impact was fewer entries (43.5%), mainly where pastoralists, farmers and other growers were unable to attend the event due to time limitations, or through lack of crops or stock for display or competitions. A quarter of festival organisers (26%) found that it had been harder to gain sponsorship from local businesses that were also facing economic decline. There had even been an aesthetic decline which affected organisation of their event, notably because of the lack of water availability for grounds and parks. At least 15 festivals had simply been cancelled. Water was critical. It is a requirement for many sporting and agricultural events, where fields and tracks need to be watered for safety of competitors and stock. Some festivals with water-based activities, such as fishing, found the condition of lakes and rivers inadequate. Yet droughts raised environmental awareness and as many as 43% of organisers believed their event played a role in helping their community adapt. Festivals lifted community spirit and brought the community together for happier times, however briefly. Far beyond any economic value festivals were invaluable for community building and social networking.

Conclusion

Rural festivals have proliferated in recent years. This is a sign that the sector is growing and becoming more sophisticated, that the general public increasingly see festivals as a fun way to use their leisure time and that councils are willing to sponsor them. Many simply celebrate place, hence the many 'community' festivals in rural Australia, and promoting and showcasing place are consistent main aims. Whether places are celebrated through agriculture, the 'old-fashioned way', music or innovative cuisine, is probably relatively unimportant for community building. The main objective is coming together. Hence the Devonport Jazz Festival in Tasmania, while obviously being about jazz, also incorporated 'elements of education, food, visual art, dance and film into the traditional jazz festival program'. While in some cases festivals articulate important goals of reconciliation, sustainability or challenging negative stereotypes, rarely do they do so without music, food, dancing or socialising. In some cases there is controversy when an event smacks of something problematic – drunken Elvises, 'different' subcultures, noisy utes, heavy metal music – that may literally be disruptive (noise, vomit, environmental degradation) or challenge (some) people's conservative perspectives of what country life should be (straight, white, quiet, etc.). In such instances, 'politics' intrudes into the rural festivals scene. But challenging and even disruptive festivals are exceptions. Controversy rarely overwhelms pleasure and festivals are primarily about supporting the community as it is. Beyond social goals there are economic benefits (evident in employment and income generation) and opportunities to develop place marketing strategies.

Some festival organisers feared that there was a 'limit' to the endless proliferation of festivals and that eventually festivals would start to fail as communities became 'festivalled-out' and competition became more fearsome. This was particularly so for music festivals, where growth in numbers has not been matched by either a growth in audiences or in the number of high-quality local acts. With increasing numbers of festivals vying to secure the same number of available bands, competition becomes fiercer and the risks of failure increase. Likewise when towns become well-known for festivals, as in Byron Bay and Daylesford, there is a risk that locals will resent repeated crowds, parking problems and noise. If not carefully managed, too many festivals can reduce local quality of life; in the case of the Splendour in the Grass festival it threatened to move out of town altogether (see Chapter 6). Conversely places can lose festivals that seemed wedded to them (as with Scone's Metal Stock, which was bought out and moved to Sydney), or can take festivals for granted, hence Tweed Heads' shock after losing its lucrative Wintersun 50s Nostalgia Festival.

Festivals succeed when there are clear visions of their aims and purpose; a sense of difference and uniqueness; and meaningful engagement with audiences, seeing them as communities of interest, rather than more cynically as 'bums on seats'. Product differentiation is invaluable, with communities of interest around the most specific themes and arcane passions. Within just one festival niche – car shows – there are further niches: vintage cars, custom-designed cars, hot rods, car clubs, brand-loyal car owners (Minis, MGs, Volvos ... .) all of which have their own festivals in an increasingly organised calendar. The extent of possible diversification is endless and the market is infinite. However, local demography provides a limit to the number of festivals; too many festivals can overwhelm small places with limited capacity and few volunteers. But risks sometimes need to be taken. Weird can be good, because quirkiness gains national publicity and reputation. Blues, jazz, food and wine festivals are comparatively 'safe', and are what communities often want from local festivals showcasing local products, but they have proliferated in recent years. Ambitious organisers need something different to put their town 'on the map'.

Yet, in the end, while place marketing might be possible, most festival organisers have no interest in branding, but in local, more modest pleasures. Indeed, the real 'success stories' of rural festivals in Australia are probably not the few obvious examples of place-branding, as at Parkes or Deniliquin, but in the very large numbers of viable and lively festivals that take place every weekend across rural Australia, and celebrate some version of local community and identity. The accumulation and diversity of modest, prosaic festivals, some very specific, others largely generic, is what makes them significant. Ordinary and uninspiring they may seem but they are invaluable – even if fleeting and unassuming – to the people those run them and take part in them. Rural festivals are sites where daily social life is transformed (albeit often gently so) as the streets are taken over by gatherings of runners, sheep, cyclists, Elvis impersonators, political activists, or a staged parade, and parks become spaces of sanctioned tolerance for excess drinking, eating and partying. The best festivals breathe a sense of life and economic vitality into what can sometimes be moribund places. They build communities out of audiences, turn residents into performers and friends, shape identity and mould experiences of rural Australia.

References

Bakhtin, M. (1984) *Rabelais and his World* (H. Iswolsky, trans.). Bloomington, IN: Indiana University Press.
Burnley, I. and Murphy, P. (2004) *Sea Change: Movement from Metropolitan to Arcadian Australia*. Sydney: UNSW Press.
Connell, J. and McManus, P. (2011) *Rural Revival? Place Marketing, Tree Change and Regional Migration in Australia*. Aldershot: Ashgate.
Connell, J. and Rugendyke, B. (2010) Creating an authentic tourist site? The Australian Standing Stones, Glen Innes. *Australian Geographer* 41, 87–100.
Eaton, D. (2007) Organiser of the Tenterfield Food and Wine Show. Interviewed by Chris Gibson and Elyse Stanes, Tenterfield, August 2007.
Falassi, A. (1987) Festival: Definitions and morphology. In A. Falassi (ed.) *Time Out of Time. Essays on the Festival* (pp. 1–10). Albuquerque: University of New Mexico Press.
Getz, D. (1991) *Festivals, Special Events and Tourism*. New York: Van Nostrand Reinhold.
Getz, D. (2007) *Event Studies: Theory, Research and Policy for Planned Events*. Oxford: Butterworth-Heinemann.
Gibson, C. (2007) Music festivals: Transformations in non-metropolitan places, and in creative work. *Media International Australia Incorporating Culture and Policy* 123, 65–81.
Gibson, C. (2010) Creative geographies: Tales from the 'margins'. *Australian Geographer* 41, 1–10.
Gibson, C. and Davidson, D. (2004) Tamworth, Australia's 'country music capital': Place marketing, rural narratives and resident reactions. *Journal of Rural Studies* 20, 387–404.
Gibson, C. and Connell, J. (forthcoming) *Music Festivals and Regional Development*. Aldershot: Ashgate.
Gibson, C., Dufty, R., Phillips, S. and Smith, H. (2008) Counter-geographies: The campaign to prevent closure of agricultural research stations in New South Wales, Australia. *Journal of Rural Studies* 24, 351–366.

Gibson, C., Brennan-Horley, C. and Walmsley, J. (2009) Mapping vernacular creativity: The extent and diversity of rural festivals in Australia. In T. Edensor, D. Leslie, S. Millington and N. Rantisi (eds) *Spaces of Vernacular Creativity: Rethinking the Cultural Economy* (pp. 89–105). London: Routledge.

Gibson, C., Waitt, G., Walmsley, J. and Connell, J. (2010) Cultural festivals and economic development in regional Australia. *Journal of Planning Education and Research* 29, 280–293.

Hicks, J. (2003) *Australian Cowboys, Roughriders and Rodeos*. Rockhampton: Central Queensland University Press.

Chapter 2
Histories of Agricultural Shows and Rural Festivals in Australia

K. DARIAN-SMITH

Introduction

From the early colonial period in Australia, agricultural and pastoral (A&P) societies were central to the promotion of new farming techniques and the encouragement of productive settlement. But the annual shows organised by these societies have always been far more than farming demonstrations. For almost two centuries, rural shows have offered a blend of entertainment, commerce, education and entertainment in a unique whole-of-town festival. Throughout their complex history, agricultural shows have often been proclaimed to represent the heart of the community and the values of non-metropolitan life. Many shows have experienced considerable pressures since the 1980s in the broader context of a shifting rural economy, and changing patterns of leisure and consumerism. Despite this, hundreds of shows remain throughout rural Australia, and with the inclusion of attendance figures for the large 'Royal' shows located in capitals cities, more Australians continue to visit an agricultural show each year than any other single event.

Agricultural shows are not, however, the only rural festival in Australia with a history that stretches back to the colonial period. Other gatherings, including annual picnics, balls and race meetings were a regular feature of country life. During the 19th century, non-metropolitan communities celebrated dates of imperial and national importance with considerable enthusiasm. From the 1920s, the phenomenon of 'back-to' festivals or home town reunions commemorated the values of pioneering settlement, and served to temporarily boost local economies in the face of rural population decline. In the post-World War II decades, growing diversity in the thematic focus of new rural festivals catered to both local community interests and the needs of increased numbers of domestic tourists.

However, studies of rural festivals in Australia have been primarily concerned with issues of contemporary significance rather than this historical dimension. This is not surprising: a recent report on thousands of festivals held in rural and regional New South Wales (NSW), Victoria and Tasmania indicates that only 24% had originated more than 30 years ago, and of these, most were agricultural shows and horticultural festivals (Gibson & Stewart, 2009: 14). The history of earlier rural festivals is bound up with the patterns of agricultural, industrial and manufacturing work. Agricultural shows, for instance, coincided with the cycle of the harvest and other farming tasks. Another impetus for the development of rural festivals has been domestic tourism. By the second half of the 19th century, Australian trade unions were, ahead of many other industrialised nations, successful in securing the right of workers and their families to an annual holiday (White, 2005: 55–60). With compulsory education acts, school breaks became increasingly associated, across social classes, with the family holiday. The rapid expansion of the railways provided relatively cheap travel from the cities to emerging resorts on the coasts or mountains (Davidson & Spearritt, 2000), encouraging city people to participate in country life. Indeed, in the country towns of colonial Australia, the local agricultural show exemplified many of the characteristics that are associated with early 21st-century rural festivals, including a distinctive sense of community and place (Derrett, 2003), and opportunities for commercial activity.

Agricultural Shows in Australia

In 1813, at the small outpost of Parramatta in the British penal colony of NSW, the first public market was held for the sale of livestock. By 1822 the Agricultural Society of NSW was established, organising regular display and judging of agricultural exhibits, with medals given to winning entrants (Fletcher, 1988: 13–28). Just months earlier, the Van Diemen's Land Agricultural Society had been founded in Hobart, with the same aim of rewarding agricultural excellence and organising an annual exhibition of livestock. Agricultural societies were soon formed in Perth (1829), Adelaide (1839) and Melbourne (1840). Initially, these were restricted associations which served the interests of the farming gentry, but their membership was soon broadened to serve rural communities more generally. From early cattle fairs and ploughing matches, agricultural shows rapidly became multi-faceted events featuring commercial, government and entertainment activities.

A&P societies in Australia were based upon, and remained in communication with, the British agricultural societies (including the Royal Agricultural Society) that emerged in the wake of the agricultural revolution of the late 18th century. British agricultural societies were engaged in the promotion of progressive farming practices, drawing upon new scientific experimentation. This was particularly important in the context of European imperialism, where the different climatic and geographic features of settler colonies challenged established 'Old World' assumptions about farming. Agricultural societies played a key role in the development of primary industry in the British settler colonies of Canada, New Zealand and South Africa. In the United States, the first 'modern agricultural fair' was held in Pittsfield, Massachusetts in 1811 (Hokanson & Kratz, 2008: 16). By the late 19th century, there was an established network of state and county agricultural societies and their fairs, particularly in the north-east and mid-west regions of the United States, that were instrumental in supporting farming endeavours and the social and economic bonds of 'everyday' rural life (McCarry, 1997; Prosterman, 1995).

In Australia, agricultural societies fostered agricultural education through publications, demonstrations and model farms, and worked closely with related breed and horticultural bodies. They were strongly supported by colonial/state governments, who used the societies to promote such schemes as closer settlement and irrigation. The most important activity of each agricultural society was the organisation of an annual show, with competitive judging for livestock, produce and manufacturing (Figure 2.1).

As the pastoral industry developed, agricultural and mining production shifted further inland, and rail and other transport networks spread, and new rural towns were established in the Australian colonies. Within months of being officially gazetted as a town, many boasted their own agricultural society and had laid aside land for a showground. Almost 70 new agricultural societies were established during the 1880s alone, a decade of vigorous rural expansion in Victoria, NSW and Queensland. By Federation in 1901, some 1000 shows were held in a seasonal cycle each year throughout regional Australia.

From their humble beginnings, agricultural shows provided increasingly elaborate exhibitions of local and colonial wealth. By the 1870s, the goods on competitive display encompassed agricultural machinery, animals, unprocessed and processed produce, industrial and manufactured wares, flowers, technological inventions and, increasingly, domestic arts and crafts and displays of human prowess from bush skills like woodchopping to tent boxing. These comprehensive interests were evident in the adoption of terms such as 'horticultural' or 'industrial' in the names of

Figure 2.1 'Cattle parade at the Kenilworth Showgrounds', ca. 1950. Image courtesy of the Sunshine Coast Libraries

societies. For instance, in 1895 in the Victorian Mallee region, the Hopetoun District Agricultural, Horticultural, Floricultural, Pastoral and Field Trial Society formed, soon becoming the town's 'premier' institution and its show 'the main event on the calendar' (Conway, 2006: 21–25).

Agricultural shows quickly became a venue for government agencies and commercial manufacturers to promote their products, policies and ideas, and the educative function of shows cannot be underestimated (Figure 2.2). This extended to ways generations of rural people 'learnt through looking' at the 'out-of-ordinary' spectacle at the show. Showgrounds were spatially laid out around a central arena, with surrounding livestock and commercial pavilions, and an area for popular entertainment known as 'sideshow alley'. Carnival operators or 'showies' followed the circuit of show dates through regional Australia by train or wagon, initially displaying their acts at 'night shows'. The Showmen's Guild of Australasia was formed in 1909, and by this time games and rides were held during the day. But the sideshows also provided a 'window on other worlds' for isolated rural communities, with displays of people from other places, human bodily difference such as the 'Tallest Man' or the 'Bearded Lady' and explicit sexuality in acts like 'Vanessa the Undresser'. Such

Figure 2.2 'BHP Exhibition at Newcastle Show', 1940. Image courtesy of the Newcastle Region Library

entertainments provided a 'vital site for many Australians to see human performances of otherness' (Broome & Jackomos, 1998: 55; see also Broome, 2009) and added considerably to the experience of the show.

By the late 19th century, agricultural shows had adopted 'traditions', including the 'grand parade' of winners of all competitive categories in the main arena of the showground. Elaborate arrangements of produce encouraged competition between towns and regions (Figures 2.3 and 2.4). The major city shows, such as the Royal Easter Show in Sydney and the Brisbane Exhibition or 'Ekka', featured colonial/state-based competition through breath-taking District Exhibits of primary produce (Scott & Laurie, 2008). Informal and formal networks between rural agricultural societies, colonial/state 'peak' societies and those in Britain (and increasingly the United States) encouraged a sense of farming as a global and cooperative pursuit, conceptually challenging the realities of isolation in rural towns.

In the capital cities, the 'Royal' or statewide agricultural societies managed large showgrounds and held shows that lasted for up to two weeks. These events were attended by thousands of non-metropolitan and urban

Figure 2.3 'Agricultural display at the Bundaberg show', 1912. Image courtesy of the State Library of Queensland

dwellers. They brought 'the country and city' together in a festival that emphasised social progress, national pride and imperial conquest over the Australian environment and its Indigenous peoples. In this sense, the agricultural show, whether in a remote town or a thriving metropolis, embodied and represented the ongoing process of triumphant colonialism. But the agricultural show also, as Kay Anderson has argued, 'enacts in thoroughly ritualistic fashion a triumphal narrative of human ingenuity over the non-human world', and as such can be seen as a spectacle of modernity as much as colonialism (Anderson, 2003: 423). As the pre-eminent rural festival, the agricultural show of colonial and early 20th century Australia celebrated local, imperial and national values.

Festivities of Empire and Nation

Alongside agricultural shows, settler communities also participated in other recurring or annual forms of celebration. Sporting competitions, for instance, provided opportunities for rural communities to develop a shared sense of identity across social classes. By 1850, there were some 45 reported race meetings in rural NSW. The popularity of picnic race

Figure 2.4 'Newcastle show: agricultural exhibit', *Newcastle Morning Herald*, 1951. Image courtesy of the Newcastle Region Library

meetings expanded throughout the 19th century, with improved rail connections enabling city spectators and bookmakers to attend country events (Waterhouse, 2005: 133–144). The railways also facilitated the extensive travel of entertainers such as minstrel troupes, vaudeville and circus companies into regular rural circuits, which sometimes were scheduled to coincide with the local agricultural show.

Rural communities, like their urban counterparts, celebrated dates of imperial and national significance. Queen Victoria's Diamond Jubilee in 1897, for instance, provided an occasion to commemorate both the longevity of the monarch's reign and the sweeping reach of the British Empire. Dinners, concerts, church services and displays of fireworks were held throughout the Australian colonies, including in regional towns (Irving, 1999: 8). The *fin de siècle* celebrations that ushered in Federation in 1901 were even more elaborate, providing opportunities to reflect on such national achievements as the role of pastoralism, crops and mining in the generation of Australian national wealth. The rural population were particularly enthusiastic supporters of Federation. Although many country people were lured to the cities to participate in metropolitan celebrations, many towns held their own parades, sporting competition and 'monster' picnics to mark the formation of the Commonwealth (Irving, 1999: 14).

While the Diamond Jubilee and Federation were one-off events, other notable dates were accompanied by annual holidays from school and employment. These included Christmas, Easter and New Year's Day. Labour Day was taken as a holiday in various colonies to commemorate the rolling success of the eight-hour day campaign. Empire Day, held on 24 May, was observed from the early 20th century, fostering loyalty to the British Empire. School children sang patriotic songs and participated in parades, and the day ended with a bonfire and fireworks (Firth & Hoorn, 1979). Anniversary Day, later known as Australia Day, commemorated the arrival of the First Fleet into Sydney Cove on 26 January and was widely observed across Australia (though especially within NSW) as a civic festival of nationhood – although for Aboriginal people, it was known as a day of mourning.[1]

By the 1920s, in the wake of World War I, the most significant annual event in any country town – as in the nation as a whole – was Anzac Day. Although the disastrous landing of Australian troops at Gallipoli on 25 April 1915 had been commemorated as early as 1916, it was not until 1927 that Anzac Day was gazetted as a national holiday in all Australian states. In the interwar years, with little exception, rural communities raised the funds for a war memorial to be erected in a prominent position in their town. Such memorials, commonly in the form of a stature or a plinth, were revealed on a dedicated 'unveiling day' which according to Ken Inglis, 'had resemblances to such civic festivals as Empire Day or Anniversary day, and to patriotic rallies ... held during the war' (Inglis, 1998: 209). By the 1930s, the rituals performed on Anzac Day were firmly established as 'traditions': the dawn service, the veterans' march, two-up gambling games, and the organisation of ex-servicemen's reunions (Inglis, 1998).

It is perhaps problematic to consider Anzac Day as a 'festival', for it was a day of mourning for the war dead. Indeed, Anzac Day was observed in the interwar period, and again in the decades following World War II, through individual and collective expressions of both patriotism and loss. The significance of Anzac Day within Australia as the most 'sacred day' of the year was acutely acknowledged in country areas, where close knit communities felt the absences of local men who had been killed and wounded in the two world wars. But for much of the 20th century, in terms of community participation across class and generation, Anzac Day and its associated civic activities remained an important annual event in rural and national life.

Between the Wars: Festivals of Pioneer Life

During the 1920s, and concurrent with the rise of Anzac Day, a new form of local festival emerged. This was the 'back-to' festival, where those who had migrated to the cities in previous decades were invited back for a weekend or even a week of civic activities honouring the pioneers of European settlement (Figure 2.5). As the 'come-backs' mingled with local residents at dances, church services, smoking 'socials' and cricket games, they were given the opportunity to experience, yet again, the social interactions associated with the 'old-fashioned friendliness and hospitality of small town life' (Davison, 2000: 204).

Back-to events emphasised the historic and ongoing ties between country towns and their former residents, recognising that permanent return was not a viable option for those who had, from necessity, moved elsewhere. These festivals were most prevalent in Victoria, where they were organised at major regional centres such as Bendigo and Ballarat as well as at smaller agricultural and, in particular, mining (and ex-mining) settlements. Helen Doyle has estimated that between 1917 and 1939, there were a staggering 239 back-to festivals in rural Victoria, with only 51 in NSW and 28 in South Australia (Doyle, 2005: 244). The considerable effort involved in their organisation meant that while some towns held back-to festivals every couple of years, the majority could only manage to do so once in a decade. Aided by the efforts of city-based groups of former residents and ex-pupils

Figure 2.5 'Back to Moonta' celebrations', 1927. Image courtesy of the State Library of South Australia

of country schools, the back-to festival encouraged nostalgia for a past associated with the places of childhood and young adult life. They also provided a temporary stimulus to the economy of many small rural towns, and the souvenir booklets and special editions of local newspapers attest to the commercial interests of local traders. Nearby attractions and scenic spots were promoted to the visitors, contributing to local tourism.

Perrie Ballantyne's poetic study of 'the cultures of remembrance' in South Australia's Willochra Plain, situates the 'back-to' festivals of the interwar period as held 'against a backdrop of drought and regional abandonment' (Ballantyne, 2004: 58). For one week in September, the population of the town of Quorn would double as dispersed former residents arrived by rail, bringing the struggling settlement temporarily 'back to life' (Ballantyne, 2004: 59). The festival honoured the town's pioneers, with speeches from the mayor and other leading citizens harking back to the difficult days of the first European settlement. The optimism of an earlier period when agricultural expansion into the Willochra Plain falsely promised bumper harvests of wheat were momentarily reignited. As Ballantyne (2004: 70) points out, ' "Comebacks" were sometimes "pioneers" who had left the region to try their luck on land elsewhere in the country, but the fact that their commitment to this place had not been enduring did not appear to weaken their status as figures around which the community could weave its story of origins'.

Indeed, in rural towns around Australia there were, in the first decades of the 20th century, older men and women still alive from the first generation of settlers. 'Back-to' festivals offered the chance for younger generations to reconnect with these people, and a sense of a living past was evident in the re-enactments of key historical moments that were a feature of many commemorative programs. The triumph of white Britishness was a key element in such history-telling; while the presence of Aboriginal people as prior inhabitants of the land was acknowledged, it was depicted in terms of their 'friendliness' to the white settlers rather that their dispossession (Doyle, 2005: 268).

The phenomenon of back-to festivals was symptomatic of a particular moment in inter-war Australia. They highlighted the 'hometown pilgrimages of soldiers and come-backs' alongside explicit reference to 'the pioneering journeys of the early settlers' (Doyle, 2005: 252). They also reflected the national concern about population drift from rural areas as a result of extreme climatic conditions and the economic crisis in primary industries. By the late 1940s and 1950s, back-to commemorations dwindled away, to be replaced by festivals that were firmly focused on a modern future for rural Australian communities.

Celebrating the Modern

During World War II, and most particularly by 1942 when Australian troops were engaged in the Pacific military theatre, there was a halt to community festivities not associated with the war effort. Voluntary labour was comprehensively directed towards providing comforts for service personnel, civil defence activities and war-related fundraising. The federal War Precautions Act introduced restrictions on services and the rationing of goods, including petrol, and military authorities requisitioned showgrounds for wartime usage. Many agricultural societies, from those in small regional towns to the peak city 'Royal' societies, disbanded their activities for all or part of the duration of war.

In the decades after World War II, most festivals in country towns continued to be associated with agricultural activities. The federal program of postwar reconstruction strongly emphasised the ideology of decentralisation away from the capital cities, the need to expand the rural economy, and the importance of providing adequate educational, commercial, health and recreational services to rural towns to arrest population decline and even attract new residents. Social progress was seen as inevitable and planning schemes for suburban developments in cities and country towns emphasised the importance of community participation, successful family life and national growth. After decades of economic depression and war, the 1950s and 1960s were a period of relative prosperity for the rural sector. This was evident in the 'long boom' in export prices for primary products such as wool and wheat, the rapid expansion of mining ventures and improved technologies in farming, communications and transport.

In rural areas, the organisation of agricultural shows and other festivals remained dependent on volunteer labour and the contributions of local businesses. There were growing commercial interests in the promotion of rural attractions, including festivals, to wider audiences drawn from the city. Tourism was aided by the increased mobility of the postwar Australian population. Car ownership was on the rise: about 20% of the Australian population owned a car in 1953, and 35% just a decade later (Davison, 2004).

More people were travelling from the cities to country areas for holidays, and such visits were often timed to coincide with shows and other seasonal events. By the 1950s, road travel had almost completely displaced the special passenger trains that had, from the late 19th century, carried visitors to agricultural shows. Sideshow operators and travelling circus troupes also abandoned the railways for road transport in their annual circuits throughout regional Australia.

Although only a handful of new agricultural societies were established in the second half of the 20th century, existing shows experienced a renewed sense of purpose and growth in the scope of competitive activities, entertainment and commercial transactions. The boom in sales of mechanised farm machinery, for instance, was reflected in the copious space devoted to their display (although by the 1970s the trend was underway for separate 'field days' for demonstrations of modern farming equipment). Government and farming bodies were particularly eager to use shows to promote scientific farming as well as providing information on trade and social policies. Among the commercial goods on display were domestic appliances, televisions, and there was an increased emphasis on the range of showbags (or sample bags) for sale. Entertainment became more 'spectacular' with displays of fireworks, rodeo and driving events, and concerts and cultural performances. There was also increased encouragement for young rural men and women to participate in events that highlighted modern country life. One example of this was the evolution of 'Miss Showgirl' events from localised fundraisers to 'sophisticated' and 'cosmopolitan' regional and ultimately national competitions that, by the 1960s, had much in common with beauty pageants (Darian-Smith & Wills, 2001).

In parallel with the expansion of agricultural shows, there was a new diversity in other non-metropolitan festivals in the post-World War II decades. Three distinctive rural festivals that emerged – the Shakespeare Festival at Swan Hill, the Festival of the Snows at Cooma and the Darwin Festival – are illustrative of how broader changes within Australian society were embraced at a local level. They are indicative, too, of a new desire to celebrate the spirit of a progressive nationhood, which was exemplified through a heightened and increasingly cosmopolitan embodiment of modern citizenship.

In 1947, the small town of Swan Hill, situated in north-eastern Victoria on the Murray River and with a population of around 5000, launched an annual Shakespeare Festival with a procession of thematically decked floats, and the staging of highlights from *Romeo and Juliet*. The festival was championed by Marjorie McLeod, a formidable woman who was the author of several social realist plays about Australian working-class and rural life, and was the co-founder of the Melbourne-based National Theatre. McLeod persuaded the recently formed Swan Hill National Theatre to make productions of Shakespeare its speciality. Under her guidance, the Shakespeare Festival grew into a week-long event and included a 'street procession, a Shakespearean play, an experimental theatre, playreading and an arts ball' (Feldtmann, 1973: 149). Local businesses promoted the Festival through window displays; the town council provided financial and other support. There were capacity audiences for the theatrical productions held in the Memorial Hall and outdoors in Riverside Park (Figure 2.6). A succession of governors of Victoria, members of the British Council and visiting actors from the Royal Shakespeare Company were invited to open the Festival. By the 1950s, the conservative Prime Minister Robert Menzies, a man of traditional cultural tastes, had become its official patron (Gaby, 2007: 170).

The success of Swan Hill's Shakespeare Festival is extraordinary on many levels. It has been the only major festival devoted to Shakespeare ever organised in Australia, although a Shakespeare play was the occasional centrepiece of the professional arts festivals beginning in Perth and in Adelaide in the 1950s and 1960s (Gaby, 2007: 170–171). The Swan Hill Festival also predated the Stratford Shakespearean Festival of Canada, founded in the town of Stratford in south-western Ontario in 1952 and which was to become North America's leading non-metropolitan repertory company.

Figure 2.6 'Shakespeare Festival', 1959. National Archives of Australia

In the postwar decades, for Australia – as for Canada and elsewhere in the wider Anglophone world – the performance of Shakespeare's plays was seen as desirable cultural capital. This was just part of a broad emphasis in Australia on the value of 'good' reading, especially the literary classics, in the development of an informed citizenship. Reporting on the Swan Hill Shakespeare Festival in April 1952, the *Australian Women's Weekly* stated that the town council was 'keen for our cultural progress to keep abreast of our material riches' (quoted in Buckridge, 2006: 363). There is no doubt that Swan Hill was a particularly forward-looking town. The town council had in 1935 funded an imposing Town Hall built in modernist style, and in 1940 it commissioned an ambitious planning scheme that drew on the latest international principles of urban design (Darian-Smith & Nichols, 2010). Within these national and local contexts, the Shakespeare Festival created a distinctive cultural identity for Swan Hill that was simultaneously rural and cosmopolitan, but above all was progressive in outlook. Indeed, the Festival's staging of Shakespeare in Australian settings and contemporary dress during the 1950s was considered artistically radical, and can be seen as indicative of this modern approach.

Nonetheless, there were contradictions inherent in the Shakespeare Festival's blend of high culture with populist community celebration. On the one hand, the Festival symbolised the renewed sentimental alignment felt by many Australians with the institutions and culture of Britain. However, the Festival also contributed to localised community bonding, with the Melbourne newspaper *The Argus* commenting in 1951 that 'Instead of being an excuse for vivid costuming and violent enunciation by an amateur company, Shakespeare in Swan Hill is the cause of a nearly 100 per cent community effort' (quoted in Gaby, 2007: 169). In 1967, Swan Hill was awarded a score of 97% in the cultural activities section of the Victoria's Premier Town competition. But by this time, the town's cultural focus was moving away from Shakespeare to its own history of settler exploration, and the 'romance' of its former river port. By the mid-1970s, at a time of a new nationalist sentiment, the Shakespeare Festival was superseded by the Swan Hill Pioneer Festival.

A second extraordinary rural festival came into existence in 1957, in the NSW town of Cooma, nestled in the foothills of the Snowy Mountains. In 1949, the federal Labor Chifley government launched the Snowy Mountains Hydro-Electric Scheme, a potent symbol of postwar nation-building effort. As Australia's largest engineering feat – and one of international significance – the Scheme diverted the waters of the Snowy River to provide power for the nation's expanding industrial base. Around 70,000 recent migrants, many who had arrived from Europe as 'displaced persons' after World War II, were employed in the construction of dams, roads, power plants and other facilities.

A Festival of the Snows was initially a relatively small-scale community event. But it is the Third Festival of the Snows in 1959 – which marked the 10th anniversary of the Snowy Hydro-Electric Scheme – that is notable. To support the Festival's growth, major infrastructure improvements included an open stage or 'music shell' constructed in Cooma's Centennial Park. An International Avenue of Flags was erected around the boundary of the park to celebrate the various nationalities of the workers on the Snowy Mountains Scheme, and the festival's celebrations also highlighted this cultural diversity.[2] The Festival of Snows in Cooma can thus be positioned as Australia's first self-consciously modern 'multicultural festival', signalling how the ethnic composition of Australia was rapidly changing in the postwar era – not just in the cities, but also in some rural towns.

Finally, a third example of new models of festivals emerging in the postwar period can be found in Darwin, the remote administrative centre for the Northern Territory. In 1964, with funding from the newly formed Tourism Board, the first Darwin Festival was launched.[3] Its highlights included a showing of David Attenborough's ethnographic film 'Artists in Arnhem Land', and exhibitions of Aboriginal bark paintings and watercolours, with the latter displaying the last work of the well-known Aboriginal artist, Albert Namatjira. By 1969, to mark the centenary of Darwin's foundation as a settlement, festival events were programmed throughout the year and included a game fishing competition, the launch of an official history of Darwin and a vice-regal visit from the Duke of Kent (Berzins, 2007: 140). The Darwin festivals of the 1960s featured 'open days' at nearby government-run Aboriginal communities as an official part of their program (Berzins, 2007: 170), thus catering to the curiosity of white residents and visiting tourists about Indigenous culture.

The Darwin Festival exemplified the growing role of official tourism authorities in the funding and organisation of festivals. During the 1950s and 1960s the Northern Territory, most notably at Alice Springs and Uluru but also in Darwin and the 'top end', was opening up to tourist groups interested in its unique environmental features and Indigenous cultures. Throughout rural Australia, however, the link between rural festivals – from the 'outback' experience of the Darwin Festival to the high culture of Swan Hill's Shakespeare Festival – and tourism was to become increasingly explicit during in the last decades of the 20th century.

Conclusion

Studies of contemporary rural festivals recognise the complex interplay between the aspects of cultural expression that are generated and the immediate and cumulative economic benefits that may result for a community (Gibson et al., 2010). By 1973, what was known as 'community arts' – a flexible category of grass-roots cultural production that incorporated both the organisation of local festivals and the activities held at them – was formally established as a funding program of the federal arts body, the Australia Council (Hawkins, 1993). This government support was to assist the establishment of new rural festivals that were predominantly expressions of local identity. However, by the 1990s as rural crisis and depopulation affected many country towns, one of the solutions often touted to stimulate the local economy was the introduction of either an ongoing tourist attraction, such as the 'sound and light' historical re-enactment put on by local townspeople at the small town of Harrow in Victoria's Western District (Darian-Smith, 2002), or a local festival that would attract thousands of visitors.

Certainly, from the 1980s a spurt of new cultural festivals in rural Australia were arguably more focused on attracting external income generated by tourism than on building local community bonds. Perhaps most noticeable has been the proliferation of popular arts and music festivals. The rural town of Tamworth, NSW, has been successfully transformed into a tourist destination as the 'country music capital' of Australia with a festival that invokes the meaning of 'country' in both globalised and localised ways (Gibson & Davidson, 2004). More recently, the smaller NSW town of Parkes has sought to arrest economic decline through its reinvention as the site of the annual Elvis Revival Festival (Brennan-Horley et al., 2007; also Chapter 11).

These new festivals in rural Australia have been formed at a time when many rural agricultural shows are experiencing a period of uncertainty. A survey of more than 600 agricultural societies indicated the extent of these challenges (Darian-Smith & Wills, 1999). Agricultural societies believed that it was increasingly difficul to obtain steady financial support from the local community, and the mandatory costs of public liability insurance placed an additional strain. The aging rural population, and the movement of younger people to the cities for education and employment, resulted in a dearth of willing voluntary workers. One consequence of this was the marked recruitment of women into positions of leadership on what was, until the 1980s, the all-male domain of show organising committees. In addition to these internal pressures, agricultural shows faced increasing external competition from other forms of entertainment. These included the internet and television, but also local events such as field days and produce markets, as well as a spate of music or special interest festivals.

Nonetheless, agricultural shows remain significant annual events for today's rural communities. 'Shows are a vital part of country life and should be retained and encouraged', wrote one show society secretary. Other representative comments in our survey included 'shows have a very important role to play as a showcase for agriculture, especially to town and city folk, and an alternative day of entertainment; and they are also an important form of cohesion for local communities' (Darian-Smith & Wills, 1999: 57). While the past two decades have seen the closure or amalgamation of a number of rural shows, many others have proved to be highly adaptable and resilient and begun to attract new audiences through refocusing on niche interests such as local gourmet produce and improving marketing strategies.

The histories of agricultural shows and other rural festivals in Australia provide insights into the formation of regional identities from the colonial period, as well as contributing to the broader field of cultural and tourism studies. But they also highlight how important the sense of a 'tradition' tied to settler farming is for the current and future identity of rural communities, or at least for the ways of belonging expressed by older residents (Darian-Smith & Wills, 1999: 58). 'Shows provide an outstanding social history of Anglo-European settlement in Australia', commented one show secretary. Another stated: 'If the shows are lost, there will be a great hole in our heritage for future generations'. Such sentiments encapsulate the contemporary significance of agricultural shows to non-metropolitan towns in Australia; a significance that draws on the ways that the 'everyday' connections between people, production and place have been confirmed at shows and other rural festivals across decades and centuries.

Notes

1. In 1938, on the 150th Anniversary of the landing of the First Fleet, Aboriginal activists including William Cooper and Pearl Gibbs organised the 'Australian Aborigines Conference: Sesqui-centenary: Day of Mourning and Protest'.
2. In 1988, federal bicentennial funding supported a curving 'timewalk', with mosaic panels recording the history of the Monaro and Cooma. In 1999, as part of the 50th Anniversary of the Snowy Mountains Scheme, the original Avenue of Flags was extended.
3. In 1978, the Bougainvillea Festival commenced to beautify the city after Cyclone Tracy, and in 1993 became the Darwin Festival, now a major arts festival.

References

Anderson, K. (2003) White natures: Sydney's Royal Agricultural Show in post-humanist perspective. *Transactions: Institute of British Geographers* 28, 422–441.

Ballantyne, P. (2004) Unsettled country: Cultures of remembrance and the ruins of South Australian colonialism. PhD thesis, University of Melbourne.

Berzins, B. (2007) *Australia's Northern Secret: Tourism in the Northern Territory 1920s–1980s*. Glebe, NSW: Baiba Berzins.

Brennan-Horley, C., Connell, J. and Gibson, C. (2007) The Parkes Elvis Revival Festival: Economic development and contested place development in rural Australia. *Geographical Research* 45, 71–84.

Broome, R. (2009) Not strictly business: Freaks and the Australian showground world. *Australian Historical Studies* 40 (3), 323–342.

Broome, R. and Jackomos, A. (1998) *Sideshow Alley*. Sydney: Allen & Unwin.

Buckridge, P. (2006) Readers and reading. In C. Munro and R. Sheahan-Bright (eds) *Paper Empires: A History of the Book in Australia 1946–2005* (pp. 362–367). St Lucia: University of Queensland Press.

Conway, K. (ed.) (2006) *The Show's Not Over: Celebrating 110 Years of Agricultural Shows in the Mallee, at Hopetoun*. Hopetoun: White Crane Press.

Darian-Smith, K. (2002) Up the country: Histories and rural communities. *Australian Historical Studies* 118, 90–99.

Darian-Smith, K. and Wills, S. (1999) *Agricultural Shows in Australia: A Survey*. Melbourne: University of Melbourne.

Darian-Smith, K. and Wills, S. (2001) From Queen of Agriculture to Miss Showgirl: Embodying rurality in twentieth-century Australia. *Journal of Australian Studies* 25 (71), 17–31.

Darian-Smith, K. and Nichols, D. (2010) Derelict Dell and Terrible Square: The plans of Frank Heath and the citizens of Swan Hill for regional development in postwar Australia. In D. Nichols, A. Hurlimann, C. Mouat and S. Pascoe (eds) *Greenfields, Brownfields, New Fields: 10th Australasian Urban History/ Planning History Conference Proceedings* (pp. 71–80). Melbourne: University of Melbourne.

Davison, G. (2000) *The Use and Abuse of Australian History*. Sydney: Allen & Unwin.

Davison, G. (2004) *Car Wars: How Cars Won our Hearts and Conquered our Cities*. Sydney: Allen & Unwin.

Davidson, J. and Spearritt, P. (2000) *Holiday Business: Tourism in Australia since 1870*. Melbourne: The Miegunyah Press.

Derrett, R. (2003) Festivals & regional destinations: How festivals demonstrate a sense of community and place. *Rural Society* 13, 35–53.

Doyle, H.W. (2005) Australia infelix: Making history in an unsettled country. PhD thesis, Monash University.

Feldtmann, A. (1973) *Swan Hill*. Adelaide: Rigby.

Firth, S. and Hoorn, J. (1979) From Empire Day to Cracker Night. In P. Spearritt and D. Walker (eds) *Australian Popular Culture* (pp. 17–38). Sydney: George Allen & Unwin.

Fletcher, B. (1988) *The Grand Parade: A History of the Royal Agricultural Society of New South Wales*. Sydney: The Royal Agricultural Society of New South Wales.

Gaby, R. (2007) An Australian Stratford? Shakespeare and the Festival. *Journal of Australian Studies* 31 (90), 167–176.

Gibson, C. and Davidson, D. (2004) Tamworth, Australia's 'country music capital': Place marketing, rurality and resident reactions. *Journal of Rural Studies* 20, 387–404.
Gibson, C. and Stewart, A. (2009) *Reinventing Rural Places: The Extent and Impact of Festivals in Rural and Regional Australia*. Wollongong: Australian Centre for Cultural Environmental Research, University of Wollongong.
Gibson, C., Waitt, G., Walmsley, J. and Connell, J. (2010) Cultural festivals and economic development in nonmetropolitan Australia. *Journal of Planning Education and Research* 29, 280–293.
Hawkins, G. (1993) *From Nimbin to Mardi Gras: Constructing Community Arts*. Sydney: Allen & Unwin.
Hokanson, D. and Kratz, C. (2008) *Purebred & Homegrown: America's County Fairs*. Madison, WI: Terrace Books.
Inglis, K.S. (1998) *Sacred Places: War Memorials in the Australian Landscape*. Melbourne: The Miegunyah Press.
Irving, H. (1999) *To Constitute a Nation: A Cultural History of Australia's Constitution*. Melbourne: Cambridge University Press.
McCarry, J. (1997) *County Fairs: Where America Meets*. Washington, DC: National Geographic Society.
Prosterman, L. (1995) *Ordinary Life, Festival Days: Aesthetics in the Midwestern County Fair*. Washington & London: Smithsonian Institute Press.
Scott, J. and Laurie, R. (2008) *Showtime: A History of the British Exhibition*. St Lucia: University of Queensland Press.
Waterhouse, R. (2005) *The Vision Splendid: A Social and Cultural History of Rural Australia*. Fremantle: Curtin University Books.
White, R. (2005) *On Holidays: A History of Getting Away in Australia*. North Melbourne: Pluto Press.

Chapter 3
Rural Festivals and Processes of Belonging

M. DUFFY and G. WAITT

Introduction

Festival activities lure us in and arouse emotions that have the potential to encourage us to be more open with others. We stop for a moment and listen, perhaps sing along with performances, smile, talk, buy trinkets, eat the local foods, and generally get caught up in the festival moment. Or, at times we may feel quite alienated by these events, rejecting the sorts of activities and performances we come across, and not wishing to be part of these activities, we grimace, cover our eyes and ears and hurry past. At a festival, then, any notion or feeling of belonging is most deeply created out of the bodily and emotive experiences of being 'in the groove together' (Keil & Feld, 1994: 167). Emotions are activated through festival activities that encourage crowd inter-mingling, such as listening to music performances, joining in dance, the aromas and tastes of food and other forms of shared experiences.

This chapter explores the ways in which rural festivals are significant events for communities as a means to enhance and foster feelings of belonging. The chapter is structured into five sections. The first section outlines why belonging has emerged as an issue in rural Australia. The second section outlines research that has adopted a representational approach to examine the role of festivals in the processes of belonging (as well as alienation and exclusion). This research highlights how festivals facilitate the expression and demonstration of particular values, cultures and histories. Festivals help sustain narratives of belonging through bringing people together to share participating in various activities, but are also an exercise in remembering the past. This is followed by a third section which examines recent research that explores the significance of different registers of belonging, specifically emotions. The fourth section is concerned with the cultural politics of emotion triggered by the sounds of the biennial Four Winds Festival, held in Bermagui, New South Wales. Attention is given to how the disparate emotions triggered by meaningful sounds of music can operate to orient people towards, or away, from articulating a sense of belonging. To conclude we underscore the importance of the cultural politics of emotion to rural festival research.

A Question of Belonging

Issues of belonging have become especially relevant in many Australian rural areas. In the context of rural Australia, public feelings of belonging (and not belonging) have been aroused by current debates about Native Title, immigration, migration and population change, climate change, and the restructuring of primary industries. For example, federal, state and local government concerns about supporting economy and development, and health services have led to immigration programs encouraging migrants to move into regional Australia. Even so, images of contemporary rural Australia are somewhat ambiguous. On the one hand, there is the popular media image of the dying country town, with those who remain struggling to make a living on the land while watching their children rapidly head for the cities and a better life (Duffy et al., 2007). The social health of rural communities is currently of concern following a range of studies that present findings of stagnation and declining standards of living, reduced lack of access to goods and services, social isolation, and problems of poor health and suicide, all of which have led to thinking about ways to reinvigorate the viability and health of rural communities (Cocklin & Alston, 2003; Dunn & Koch, 2006; Dunphy, 2009; VicHealth, 2001). On the other hand, television travel shows and glossy lifestyle magazines have us longing for the idyllic vistas of ocean views or bush retreats that, even while at a far enough distance from the pace and smog of the city, nonetheless allow us to maintain the comforts of good food, wine and coffee.

The underlying premise of much of the international literature effectively ties the benefits of festivals to their *economic* implications as tourist attractions rather than as events for residents. However, many Australian rural councils and community groups have invested substantial resources into community events – such as festivals, fairs, farmers' markets and fêtes – with the expectation that they will consolidate notions of community, improve social engagement and heighten feelings of belonging. As Gibson and Stewart's 2009 report on Australian rural festivals found, rather than being income-generating enterprises attached specifically to the tourism industry, overwhelmingly the stated aims of festivals in Victoria, New South Wales and Tasmania were most concerned with social and cultural concerns around building community or were linked to the 'pastimes, passions or pursuits of the individuals on organizing committees' (2009: 14). Nonetheless, these festivals have a significant positive impact and flow on effect on the small economies of rural towns, with an estimated total economic activity generated by rural communities in Victoria, New South Wales and Tasmania to be around $10 billion per annum (see Chapter 1). While festivals have become a popular tool in local government policies and organisational strategies for initiating economic renewal, the festivals literature also points toward their significance in adding value to everyday lives (Derrett, 2003; Duffy *et al.*, 2007; Kong & Yeoh, 1997; Quinn, 2003). Festivals are inherently about celebrating community and are understood as community-building activities. Thinking about festivals in this way highlights how they function to encourage notions of community participation, enhance local creativity and foster community well-being. Yet, these are complex, potentially divisive processes, and raise many questions. What is 'the community' celebrated at festivals? What of those who *feel* excluded from the celebrations of identity intrinsic to festivals, and who consequently do not feel part of the community? The next section turns to answer these questions.

Community and Belonging as Representative Practice

Belonging assumes identification, in terms of social location (gender, race or class), with regards to narratives of identity (Probyn, 1996), and how these various constructions are attributed ethical and political value (Yuval-Davis, 2006). Questions around who belongs to our variously imagined communities (Anderson, 1983) ask us to consider what it means to be a member of any community, and the ways in which we then articulate such meanings. Community is often understood in terms of an attachment or bond to particular people and places (Mulligan *et al.*, 2006).

Hence, it is possible to study rural festivals through a narrative or representational approach: collecting stories about active participation in creating, celebrating and engaging with ideas of collective bonds and belongings – forged either by foods, ethnicity, drinks, animals, arts, sports, ideas or activities. As such, festivals are most often understood in terms of community building activities (Gibson & Stewart, 2009). Indeed, much festival research has focused on relationships between people and place through the ways in which identity, festivals and belonging are constituted through narrative or representational practices.

Festivals offer opportunities for a locale's disparate groups to be brought together in ways that emphasise certain commonalities, albeit often centred on notions of belonging to a (local) place (Auerbach, 1991). In this way, festivals have come to be a popular platform for local government and other non-government organisations to create and encourage a sense of social cohesion in culturally diverse everyday social worlds, often within a discourse of tolerance or multiculturalism. These activities are often normative processes, for, as Lavenda suggests, festivals are about 'people celebrating themselves and their community in an "authentic" and traditional way, or at least emerging spontaneously from their homes for a community-wide expression of fellowship' (1992: 76). Hence, a community may appear to be something that simply resides in and is connected to a particular physical location. Festivals may be deployed to support such bounded notions of community and place, through markers of identity such as race, class, gender and sexuality. These markers, such as flags, posters, and banners, are representations of preferred community: they serve to regulate and reinforce notions of who belongs. So while festivals may initially seem unproblematic in their celebration of community, representative practices also ensure they may operate as spaces of exclusion as well as inclusion. Yet, overlooked in this representational approach is the importance of emotions in sustaining a sense of community and belonging. The next section examines the importance of the cultural politics of emotion in festivals research.

Community and Belonging as Emotional Attachments or Detachments

Despite the recognition of the importance of the 'big thrill' and the 'buzz' of festivals in sustaining place-based attachments (Chalip, 2006) – emotions that can perhaps be gleaned from interviews and surveys – there has been until recently far less engagement with the sensuous, embodied and emotional dimensions of festivals. Belonging is an emotional attachment, about *feeling* 'at home' and 'safe' (Yuval-Davis, 2006: 2). Festivals are therefore, significant to the emotional cultural politics of belonging: they bind people together through joy, but also may playfully question or more forcefully challenge who belongs.

Turning our attention to emotional, experiential elements also reminds us that there is something important about rurality as the context for festivals. The energies that drive individuals shaping rural and regional communities lead to very different experiences of the everyday. A study of festivals held in rural locations can uncover much about the stories, experiences and agency of emotion circulating in everyday regional life, and the dynamic and vibrant ways in which notions of community, identity, place and belonging are being experienced and related. As Robyn Mayes argues (in Chapter 10), festivals provide a crucible for different emotions to circulate. This outlook is echoed by Lyndon Terracini, founder of the Northern Rivers Performing Arts, who argues that the ways arts and cultural practices intersect in regional Australia is more than simply setting cultural activities in a particular place or region. The event of a festival, he declares,

> should be about fundamentally understanding what resonates within the people who live there, left there, or died there; and about translating those deep local associations for the benefits of a much wider audience. It should also be a place where big ideas take root, where inspirational individuals and artists who believe passionately in their cultural and artistic responsibilities can plant seeds that will grow to nourish the minds of a broader community. (Terracini, 2007: 11–12)

Emotions, how they are absorbed by individuals and transferred between bodies, therefore play a crucial part in the relationships created and built up between people, place and festivals. In short, emotions are not neutral, they are instead full of multiple implications through how they are generated, shared, shed and passed between persons.

Festivals are increasingly being understood in terms of the emotional transmissions that move through one person, or spread from one body onto/into another, then another. In other words, attention has turned to understanding how festival attendees 'soak up the atmosphere' through their various senses. Greater attention is being given to examining how emotions moving through the body cause an attitudinal perspective, or orientation, towards or away from a particular thing. For example, at a festival it may be how emotions are triggered by political affiliation, or the lyrics and music of folk songs (Chapter 15), the smell of a particular wildflower (Chapter 10), the taste of food, wine, cheese or chocolate (Chapter 1), or the intoxication of drugs and alcohol at music festivals or at Oktoberfest celebrations. In the case of SnowFest in the Hunter Valley, the pleasure from touching snow is a catalyst for moving people together (see Chapter 4). People become more alike through the transmission of joy from the sensual pleasures of snow, in a place where snow does not normally exist. Alternatively, the cultural politics of pride are mobilised through the sounds and sight of Scottish massed pipe bands. Ruting and Li (Chapter 16) investigate how the emotional transmission triggered by bagpipes at the Bundanoon Highland Festival – the 'spine-tingling sound' – may cause a stronger sense of orientation and attachment towards a particular understanding of Scottish heritage. What festivals unearth are spontaneous, instinctive emotional responses: processes that are simultaneously intuitive, neurological, cultural and social. In the next section, we discuss one case study where such instantaneity was critical to how emotional transmission occurs: through how people listened at a classical music festival, The Four Winds Festival, held at Bermagui, New South Wales. We discuss the emotions triggered by sound and how these emotions moved within and between participants, orienting participants closer to each other. This was expressed in terms of belonging and community.

Emotional Transmissions at the Four Winds Festival, Bermagui, New South Wales

Listening to the sounds of festivals may create the communication of emotion that encourages an openness to others and feelings of belonging together (Ansdell, 2004; Ehrenreich, 2007; Waitt & Duffy, 2010). Such a response is illustrated by one attendee at the Four Winds Festival, held in the coastal town of Bermagui, who explained during an interview, 'People come together as a whole instrument – made up of the many human beings and form a new living being – it is just transitory – and then it is gone'. How individuals listen to the sounds is significant to accessing how the emotions moving within festival attendees' bodies causes an attitudinal perspective towards, rather than away from, other bodies.

In our work on music festivals (Duffy *et al.*, 2007, 2010; Waitt & Duffy, 2010), we have had an opportunity to reconsider the processes of belonging offered because of the very nature of music. Sound and music have physiological and affective impacts on bodies that are immediate, that do not move through neurological pathways of cognitive thought (Levitin, 2006). Therefore, it is not simply that music *represents* identity and so notions of belonging, as some researchers have suggested. Rather, it is that subjectivities are constituted within the very unfolding of the sonic event (Jazeel, 2005). As music therapist Gary Ansdell (2004: 72) points out, 'music possesses certain qualities and 'powers' that allow personal and social things to happen', and this is significant to understanding the role of sound and music in helping to constitute our social realities. Sound brings our attention back to the flesh of our bodies; we respond to rhythms, melodies, timbres and to others likewise engaged (Benzon, 2001; Duffy *et al.*, 2007; Wood *et al.*, 2007). In other words, a focus on sound brings our attention to the intuitive, emotional, psychoanalytical processes of subjectivity (Thrift, 2008). This does pose a crucial methodological question: can we, in a meaningful way, make sense of the innate biological responses aroused by sound and music at festivals?

Nichola Wood, Susan Smith and Michelle Duffy (2007; see also Duffy & Waitt, forthcoming) considered how we could explore and experiment with what music is and how it works as music in the world. We wanted to consider approaches to research that emphasise and consider sound's peculiar qualities of ephemerality and its non-representational nature, so as to think about 'how music constitutes an elaboration of the new ... which emerges out of and in response to a whole range of material, social, cultural and economic relations' (Wood *et al.*, 2007: 869). In the Four Winds Festival study, we took up some of these auto-ethnographical approaches, including audio-video-recording and 'participant-sensing' note-taking, as a means of gaining evidence on how audience members and musicians respond to sound and music. However, while these approaches can tell us something about people's outward responses to what they are experiencing – smiles, dancing, tapping feet or grimaces, ears covered, moving quickly away – they do not tell us much about a deeper set of responses and how these come to be translated into disconnections or connections to others, and thence feelings of belonging. Nor does it allow for individual differences in how emotions may be registered, since there is an assumption that all people respond to things in the same way that we can then read and interpret. What we wanted to capture were the instantaneous emotions experienced by those present at music festivals so as to explore the process of connectedness as these emotions were transferred and passed through festival attendees at the event. We also wanted to gain some understanding of how people experienced a particular rural place; how did a festival held in their town perhaps facilitate or create an experience that addressed feelings of community connection? With these goals in mind, we designed a sound diary as a means to uncover the links and connections between an individual, his/her community and place, and how these may be enhanced or decreased by participation at a music festival.

Our research was conducted in Bermagui, a coastal town of around 2000 people, located approximately 400 km south of Sydney, and at its Four Winds Festival, a classical music festival held every second year over the Easter weekend. Bermagui is now known as a 'sea-change' town, with an influx of particularly older, relatively wealthy residents from Canberra, Melbourne and Sydney. Yet, its earlier history is one of a variety of different waves of in-migration. In the late 19th century, early European setters displaced the original Aboriginal communities of the Yuin nation, while the 20th century saw the arrival of fishers, loggers and farmers, then those of the 1960s counter-culture looking for cheap and alternative places in which to live.

The Four Winds Festival began in 1991. The main festival venue is approximately 9 km outside of Bermagui in a natural amphitheatre

(Figure 3.1). While the festival initially had an audience consisting of a handful of friends, since 2002 the biannual weekend festival draws around 1000 attendees each day to the ticketed program. In 2008 the festival organisers introduced a free-opening concert in a public park in the Bermagui township, as part of their agenda for greater social inclusion. This free concert also brought together friends and families from the surrounding towns.

In our exploration of this festival as a community-enhancing project, we wanted to capture how people came to understand themselves as part of the Bermagui community. Our use of sound diaries was designed to work specifically with an individual's personal responses and knowledge of sound, as well as their experience of and affective reactions to these sounds. The way in which sound diaries were used involved two stages. First, each participant was equipped with a small hand-held digital recorder and asked to record meaningful sounds of their everyday world. This allowed people to think about the sonic qualities of place in their daily lives: what was important to them, how it made them feel and to contemplate why these sounds were significant. A number of participants

Figure 3.1 Pre-concert rehearsal, Four Winds, Bermagui (photograph Gordon Waitt)

also included their spontaneous bodily and emotional reactions to the sounds they had recorded. Next, participants were asked to take part in a conversation soon after the recordings had been made. One of us and the participant listened to the recordings together, and we talked about what was recorded. While reviewing each sound byte, participants recaptured the sense of instantaneous exuberance and/or despair being registered, as well as the possibilities to articulate more measured, self-reflexive accounts about sounds and emotive experiences (for a fuller discussion of this methodological approach, see Duffy & Waitt, forthcoming).

When asked to make their sound diaries of their daily lives, perhaps not surprisingly, those who participated were drawn to sounds of the non-human world, and more specifically, sounds that were located within the landscape of Bermagui (the sounds of the surf, wind, birdcalls, rain, even frogs). Also, non-human sounds were significant to processes of marking out networks of belonging (car engines, the chatter in a cafe while coffee is made); the everyday, almost mundane sounds that nonetheless triggered emotional connections with notions of home. In the case of the festival, the site is carefully managed to confirm the idea of Bermagui as 'paradise', and little is allowed to disrupt the conventional pleasures of this paradise. Any distracting human sounds, such as traffic, that might intrude are banished. Indeed, for those seduced by nature as paradise the affective power and emotional rewards of the festival were heightened by the combination of sounds of bellbirds, frogs and wind moving through the trees. Such heightened responses were demonstrated in the ways participants haltingly described what it was they felt. For example one participant, Sarah, while recording the sound of the didjeridu, vividly added her own comments, 'The didj, and energy down my spine, looking across the sea of faces, more familiar people, more people I know, my community'. In this description, the didjeridu is described as having a spine-tingling 'energy', which then reminded Sarah of her connections to those around her – as she explained to herself, this is 'my community'. What was important for our methodological approach was that these responses also opened up emotional responses worked into narratives of the interconnections between place, self and sound, that were then interpreted very strongly in terms of belonging (Figure 3.2). We hear this very clearly in the words of another festival participant, Belinda, who told us,

> One of the things, that, in your experiment draws my attention to it [community], like in the performance on Friday at the oval and here, getting moved by the music. And then I look around. I think this is my community, you know, I look around and ... there is the women in the

Rural Festivals and Processes of Belonging 53

Figure 3.2 Drummers amongst audience, Four Winds, Bermagui (Photo Michelle Duffy)

local corner shop and there is the people I know. And, I just think, how lucky are we ... And, also this community has drawn this thing [together].

The music festival gave Belinda a renewed focus on ways to express what she constructs as her sense of community, in terms of an emotive experience. She looks around and, with her heightened attention to the sound and the space in which it is heard, makes links to those around her. In this framework, belonging is an affective quality in and through which she orients herself within noise, silence, vibrations and music. Consequently, she comes to feel 'in place' through the very physiological responses of the body (see also Duffy *et al.*, 2007). These conversations arising out of the sound diaries alerted us to how listening to the festival sounds in real-time – the *lived* experiences of sounds – helps bring to life the notion of community. Here, the affective qualities of sound contributed to a sense of *communitas*, expressed in terms of strengthening intimate connections to fellow festival attendees.

Even so, the unpredictability of embodied practice needs to be reiterated. Simply framing festivals through particular styles of genres, or by promoting certain cultures and lifestyle choices, is not about bringing into being some uncomplicated utopia of living with others or with difference. Understanding affective and emotional responses requires understanding the 'openness' of bodies to listening, and this requires paying attention to listening skills, level of emotional engagement and familiarity with the affordances of the sounds (De Nora, 2000), in this instance the timbre, melody, pitch and beat. However, the disparate ways that people register and engage with a festival will inevitably lead to those who feel no connection, as we hear in Elaine's sound diary, notable because she made no recordings of the Four Winds Festival. Instead, she recorded the sound of traffic on the road, the voices of her children and next-door neighbours. From the follow-up interview it became apparent she could hear the Opening Concert from her home. However, these sounds were not meaningful, and hence not recorded. As another participant, Ruth, explained:

> The Four Winds Festival didn't really hit my radar over the weekend. I was busy with the family, and we had plans to go fishing. On Friday afternoon I could hear the drums and stuff, and then we could see from our balcony all these people wandering about over the park. We hadn't a clue what they were doing. No interest in the Four Winds Festival. It's crap. They were still all hanging around the park when we headed off to the pub with friends.

Elaine's comments are important in understanding the usual promotion of festivals as bringing about social connectedness and inclusion; she has no emotional investment in the Four Winds Festival, and hence, she describes the Four Winds Festival as 'crap', and an event she never had any intentions of attending. More significantly, she could not make sense of these musical activities and their place in Bermagui. Her disdain for the event moved her away from a collective sense of belonging forged through the Four Winds Festival. Equally, her comments suggest that engagement requires a level of emotional involvement in the event, something she clearly did not desire.

Examining how people listen provides insights to the emotional engagements that help forge relationships. Attention to listening provides, as music psychologist Ian Cross (2006: 124) argues, 'an open framework for interaction'. Listening bodies bring our focus back to the transmission of emotions within and between festival attendees, not as separate and autonomous entities, but as embodied in the specific subcultural histories of Bermagui (including the Yuin people, 'hippies' and the 'sea-changers'). In this framework, belonging is an emotional quality in and through which we simultaneously orient ourselves within the affordances of sounds of a festival, and are oriented through the very physiological responses of our bodies.

Conclusion

We experience belonging in a variety of forms, from an abstract feeling about what we may think of as *our* place or community to a much more concrete notion of belonging in which we participate in activities that display our allegiances and affiliations. In thinking about rural festivals and their role in catalysing belonging, we need to consider how emotional registers operate.

Festivals are mechanisms that can help constitute individual feelings of acceptance and belonging within an imagined, collective sense of 'we', of being part of the community – although, while this is often the intent of festival organisers, such states of euphoria and intimacy may not always come to be. The transitory nature of festivals produces different and often conflicting configurations of identity, place and what it may mean to belong. The space and time of the festival is a complex site for thinking about localness and belonging but often, too, festivals celebrate connections beyond that of the locally defined community. In this way, belonging is mobile – it moves from place to place, it moves in time – and at the same time is immobile, as it is attached to particular bodies, to our actions, feelings, and our experiences. In other instances, individuals may feel actively excluded, that these are not their sounds, smells, tastes or sights of home or community. The festival is, in fact, a paradoxical thing; festival events function as a form of social integration and cohesion, while simultaneously they are sites of subversion, protest or exclusion and alienation. It is precisely this paradoxical nature that creates the festival's socio-spatial and political significance for notions of community and belonging.

References

Anderson, B. (1983) *Imagined Communities*. London: Verso.
Ansdell, G. (2004) Rethinking music and community: Theoretical perspectives in support of community music therapy. In M. Pavlicevic and G. Ansdell (eds) *Community Music Therapy* (pp. 91–113). London & Philadelphia: Jessica Kingsley Publishers.

Auerbach, S. (1991) *How to Grow a Multicultural Festival: A Handbook for California Organisations*. Los Angeles: Los Angeles Cultural Affairs Department.
Benzon, W. (2001) *Beethoven's Anvil: Music in Mind and Culture*. New York: Basic Books.
Chalip, L. (2006) The buzz of big events: Is it worth bottling? *Kenneth Myer Lecture in Arts and Entertainment Management*. Geelong: Deakin University.
Cocklin, C. and Alston, M. (2003) *Community Sustainability in Rural Australia: A Question of Capital?* Wagga: Centre for Rural Social Research.
Cross, I. (2006) Music and social being. *Musicology Australia* 28, 114–126.
De Nora, T. (2000) *Music in Everyday Life*. Cambridge: Cambridge University Press.
Derrett, R. (2003) Festivals & regional destinations: How festivals demonstrate a sense of community & place. *Rural Society* 13, 35–53.
Duffy, M. and Waitt, G. (forthcoming) Sounds of belonging: Methods of listening to place. *Aether: The Journal of Media Geography*.
Duffy, M., Waitt, G. and Gibson C. (2007) Get into the groove: The role of sound in generating a sense of belonging through street parades. *Altitude* 8. On WWW at www.thealtitudejournal.com. Accessed 27.05.10.
Duffy, M., Waitt, G., Gorman-Murray, A. and Gibson, C. (2010) Bodily rhythms: Corporeal capacities to engage with festival spaces. *Emotion, Space and Society*, (in press).
Dunn, A. and Koch, C. (2006) National directions: Regional arts. *Regional Arts Australia*. On WWW at www.regionalarts.com.au/raa1/files/RAAbook2006.pdf. Accessed 27.05.10.
Dunphy, K. (2009) Developing and revitalizing rural communities through arts and creativity: Australia. In N. Duxbury, H. Campbell, K. Dunphy, P. Overton and L. Varbanova (eds) *Developing and Revitalizing Rural Communities Through Arts and Creativity: An International Literature Review and Inventory of Resources* (pp. 1–40). Vancouver, Canada: Simon Fraser University.
Ehrenreich, B. (2007) *Dancing in the Streets: A History of Collective Joy*. London: Granta Books.
Gibson, C. and Stewart, A. (2009) *Reinventing Rural Places: The Extent and Impact of Festivals in Rural and Regional Australia*. Wollongong: University of Wollongong.
Jazeel, T. (2005) The world is sound? Geography, musicology and British-Asian soundscapes. *Area* 37, 233–241.
Keil, C. and Feld, S. (1994) *Music Grooves*. Chicago & London: University of Chicago Press.
Kong, L. and Yeoh, B. (1997) The construction of national identity through the production of ritual and spectacle. *Political Geography* 16, 213–239.
Lavenda, R. (1992) Festivals and the creation of public culture: Whose voice(s)? In I. Karp, C. Mullen Kreamer and S. Lavine (eds) *Museums and Communities: The Politics of Public Space* (pp. 76–104). Washington: Smithsonian Institute Press.
Levitin, D.J. (2006) *This is Your Brain on Music*. New York: Dutton/Penguin Group.
Mulligan, M., Humphrey, K., James, P., Scanlon, C., Smith, P. and Welch, N. (2006) *Creating Community: Celebrations, Arts and Well-being within and across Local Communities*. Melbourne: RMIT Print Services.
Probyn, E. (1996) *Outside Belongings*. New York: Routledge.
Quinn, B. (2003) Symbols, practices and mythmaking: Cultural perspectives on the Wexford Festival Opera. *Tourism Geographies* 5, 329–349.

Terracini, L. (2007) *Platform Papers 11 – A Regional State of Mind: Making Art Outside Metropolitan Australia.* Sydney: Currency House.
Thrift, N. (2008) *Non-Representational Theory: Space, Politics, Affect.* London: Routledge.
VicHealth (2001) Rural partnerships in the promotion of mental health and well-being. On WWW at http://www.vichealth.vic.gov.au/en/Resource-Centre/Publications-and-Resources/Mental-health-and-wellbeing/Mental-health-promotion/Rural-Partnerships.aspx. Accessed 27.05.10.
Waitt, G. and Duffy, M. (2010) Listening and tourist studies. *Annals of Tourism Research* 37, 457–477.
Wood, N., Duffy, M. and Smith, S.J. (2007) The art of doing (geographies of) music. *Environment and Planning D: Society and Space* 25, 867–889.
Yuval-Davis, N. (2006) Belonging and the politics of belonging. *Patterns of Prejudice* 40, 197–214.

Part 2
Nuts and Bolts: Making Festivals Happen

Chapter 4
Local Leadership and Rural Renewal through Festival Fun: The Case of SnowFest

A. DAVIES

> *Only through leadership can one truly develop and nurture culture that is adaptive to change*
> Kotter, 1998: 166

Identifying and measuring the benefits (and costs) of festivals to rural communities is not as simple as subtracting the net investment capital from the net participant expenditure. Festivals catalyse social networking, capacity building and entrepreneurial capacities. Benefits are not just achieved during the period of the festival itself, but also through the organisation process. This chapter argues that festivals are important catalysts for local leadership in rural communities. With local leadership widely recognised as the key to organisational effectiveness and successful endogenous development activities, festivals can benefit the overall well-being of communities. This chapter reviews the role of festivals as sites for building social networking, capacity and local confidence and also for fostering transformational leadership.

A short case study of the unusual SnowFest, Gloucester (in rural New South Wales) illustrates the argument and shows how transformational leadership was developed during the project. The role of transformational leadership in influencing the nature of the festival and its contribution to Gloucester's social and economic well-being is also discussed.

Transformational Leadership

To understand the role and nature of local leadership in rural economic and social adjustment processes in Australia, local leadership can be considered to be either transactional or transformational (Davies, 2007, 2009).

The transactional/transformational leadership typology was first introduced by Burns (1978) and later extensively theorised by Bass (1985, 1988), Bass and Avolio (1990, 1993, 1994) and Avolio and Bass (1999). Transactional leadership, based on the Expectancy Theory of Leadership and the Path-Goal Theory of Leadership (Egan *et al.*, 1995) refers to the exchange relationship between leaders and followers generated to meet their own interests. Transformational leadership refers to:

> the leader moving the follower beyond immediate self-interests through idealized influence (charisma), inspiration, intellectual stimulation, or individualized consideration. It elevates the follower's level of maturity and ideals as well as concerns for achievement, self-actualization, and the wellbeing of others, the organization, and society. (Bass, 1999: 11)

The transactional/transformational typology describes two higher-order leadership categories. Since Burns (1978) first published his seminal work introducing the transactional/transformation leadership typology, considerable research attention has been given to better understand transformational and transactional leadership and how it operates.

In Australia, few studies had been published on transformational leadership (Egan *et al.*, 1995; Parry, 1999). Those that have been conducted tended to focus on leaders, rather than on the leader/follower interaction, although research on leaders has revealed two findings that are important here. First, Egan *et al.* (1995) identified that leadership was not related to the organisational type and suggested that transformational leadership occurred most frequently in organisations that operated in turbulent environments. Transformational leadership was considered 'more appropriate under exceptional conditions, such as those requiring non-routine and unusually high performance, in order to prevail and be effective, such as crisis of high levels of uncertainty' (Shamir *et al.*, 1993: 589). Second, Sarros *et al.* (2008) confirmed that transformational leadership was important in stimulating a climate of innovation and promoting adaptive capacities in followers; leaders who articulated a clear vision that set high-performance expectations for followers were effective in promoting desired patterns of change in organisations.

Most leadership studies (Australian or otherwise) have centred on transformational leadership within workplaces such as multinational companies and government agencies. Very few studies have focused on transformational leadership within informal groups such as community representative committees and local sporting clubs, and especially in rural communities.

Lack of research has, however, not curtailed the development of costly leadership training programmes, and over the past 20 years, rural communities across Australia have experienced a growth in government investment in leadership training programmes specifically targeted at informal community groups. These training programmes can be anything from half-day workshops to two-year courses involving numerous weeks of intensive lectures and skill development activities. Investment in leadership development has been pursued by government as a mechanism for building the internal adaptive capacities (and thereby resilience) of rural towns. Often associated with the rolling out of neo-liberal policy, most leadership programmes target the development of transformational leadership attributes and skills. In an analysis of the effectiveness of such programmes, Haslam-McKenzie (2002) found that the Progress Rural WA programmes, which aimed to develop transformational leadership, had been reasonably successful in achieving their objectives. As Haslam-McKenzie commented:

> [the programs] enabled participants whose commitment to their communities and industries is clear, to visualise and work towards achieving networks and leadership strategies that would enable them and those around them to have some control over the change process in a collegial and collaborative environment. (Haslam-McKenzie, 2002: 31)

Haslam-McKenzie's work is a rare example of scholarly reflective analysis on rural leadership and training.

This chapter seeks to contribute further to this nascent concern about the role and nature of transformational leadership in rural communities. The discussion centres on a case study of Gloucester and the role of transformational leadership in the SnowFest festival, which provides not only an insight into the operation of transformational leadership in a rural community setting, but also highlights the value of rural festivals as sites for leadership development.

Leadership and Rural Festivals: SnowFest, Gloucester

Gloucester Shire covers 2918 square kilometres and is located in the upper Hunter region of New South Wales. Its centre is the town of Gloucester, just over 300 kilometres from Sydney. There are five small villages in the Shire area. The Shire of Gloucester had a population of 4797 persons in 2006, half of whom lived in Gloucester town. The overall size of the Shire's population has changed little since 1996, although the town did experience considerable population turnover: both the Shire and town experienced about 34% population turnover between 2001 and 2006, driven by an increase in migration to the town from other parts of New South Wales.

While the population of the Shire has remained relatively stable, the structure of the labour force has diversified. Between 1996 and 2006, industries with a declining share of total employment included agriculture, forestry and fishing, manufacturing, wholesale trade, accommodation and food services. Industries that increased their share included construction, professional, scientific and technical services, administrative and support services, public administration and safety, education and training and health care and social assistance.

Associated with the shift in the structure of the workforce was a shift in the age structure of the population. Gloucester recorded an ageing population between 1996 and 2006 with a decrease in the proportion of the population aged 0–39 and an increase in that aged 40 and over. There was a particularly sharp decrease in the proportion of the population aged 0–9 and 25–35, which indicates a trend of increased net out-migration of young families. There was also a considerable increase in the proportion of the population aged 50–64. This shift has been driven by the in-migration of semi-retired and recently retired individuals and couples.

One of the most significant events to influence the economic and social character of Gloucester in recent decades was the deregulation of the dairy industry in 2000, before which the town supported a dairy factory, which directly employed approximately 35 full-time staff. The factory was one of the largest employers in Gloucester, with a significant multiplier effect in the local economy, purchasing fresh milk from local dairy farms, with local drivers transporting the product. Following deregulation, a large number of local producers moved away from fresh milk production. Some producers opted to concentrate on beef production, some pursued specialty value-added markets and others sold their farms and moved away from the area, and in 2001, the dairy factory closed. A number of displaced workers from the dairy factory (and their families) moved away from the community to pursue employment opportunities elsewhere.

Of comparable magnitude in its social and economic effect on the community were 1998 timber industry reforms. One of the most notable losses that followed these reforms was the 1999 closure of Boral timbers, which employed 31 full-time workers. As with the dairy industry, the timber industry had a significant multiplier effect in terms of local employment.

As the community adjusted to the decline of two traditional employment sectors, and the associated out-migration, it experienced a largely unrelated trend of increasing in-migration, mostly of people aged between 50 and 64, attracted to Gloucester for its affordable land, high amenity values and accessibility from Newcastle and Sydney. This population in-migration resulted in an influx of new capital, which stimulated building activity and demand for services such as health care. A number of service-oriented businesses targeted at this new population cohort were also established, including cafes, restaurants, bookstores and gift shops.

As Gloucester experienced population change and economic restructuring, the social impact on the community was considerable. In a series of in-depth interviews with residents and community leaders, it was widely expressed that there had been a noticeable breakdown of traditional social networks and an increase in the occurrence of social isolation. Local residents also indicated that they felt the community's identity had been undermined by the changes in the dairy and timber sectors and they were unable to associate with a clear identity for the community. Indeed, during this period, participation in traditional social and community volunteer groups declined. For example, the Gloucester Agricultural Show, which was once regarded as a mainstay of the town's annual social calendar, suffered a decline in volunteer numbers and overall community participation.

In response to the population and economic changes experienced in Gloucester, local community groups, businesses and the local government developed a number of strategies to stimulate community cohesiveness and sense of place. One of these initiatives was the SnowFest festival.

SnowFest was a one-day festival, first held in Gloucester in July 2000, and which ran each July till 2003. In 2003, the festival was expanded into a two-day event with over 12,000 people attending. The festival was cancelled in 2004, with the then organisers citing unsustainable costs for organising the event, but it was again held in 2008 as an occasional event (rather than an annual festival).

The first SnowFest festival was organised by local residents on a voluntary basis. The festival was based around a theme of snow. As Gloucester very rarely receives snowfalls, a large truck transported a trailer of snow approximately 740 km from the Snowy Mountains to Gloucester. The snow was dumped in the main street of the town, which was closed to traffic for the weekend. Donated 'treasures' were hidden in the snow and children and adults were invited to play in the snow pile and hunt for them. The festival involved a range of other 'free' entertainment options for local residents and visitors including music and street performances. Local businesses were invited to participate in the event though involvement in a scarecrow competition. Businesses built scarecrows and displayed these in their stores, attracting visitors to view and judge the entries. A variety of food and craft stalls, run by local community organisations and sporting clubs, were also set up for the festival. Most visitors stayed at the festival site for the whole day visiting market stalls, participating in games and viewing live performances.

The idea for SnowFest and the initial organisational momentum for the festival came from two local residents (referred to hereafter as leaders). SnowFest was envisaged as a catalyst event for a tourism-led recovery for the community. Both leaders, who had moved to Gloucester in the 1990s, had previously witnessed the role that local efforts could contribute to tourism-based economies in other countries and contexts (further information about these two leaders has been provided in Box 4.1). They recognised that in Gloucester, community groups and individuals had limited involvement in actively developing the tourism industry, and much of the effort to develop a local tourism industry was being undertaken by local government and private businesses. The two leaders believed that for a tourism-led recovery to be successful, the economic structure and social identity of the town needed to shift. For this to occur, wide community support for, and participation in, the tourism industry was needed. The town would have to become a 'tourist town' rather than an agricultural town with a small tourism industry. The leaders believed that a novel festival, which showcased the area's tourism assets to visitors and residents, was an appropriate method for achieving this.

The two leaders began their efforts by first seeking financial and in-kind support from the local council, community groups and private businesses. The majority of organisations and businesses approached were not supportive of the concept of a festival and did not believe it could achieve the

---

Box 4.1 Introducing the founding leaders of SnowFest

Leader 1:

Mary* moved to Gloucester in the late 1990s with her husband. Prior to moving to Gloucester they had lived in Europe and Africa. Mary and her husband achieved considerable success in their careers as reporters. They were attracted to Gloucester for its small and vibrant community, the natural attractions of the region, the accessibility to Newcastle and Sydney, and affordability of the land. Mary and her husband purchased a 'hobby' farm, which they stocked with a small herd of specialty cattle. On moving to Gloucester, Mary and her husband retired from full-time work and Mary quickly became involved in

> community organisations and social clubs. Of her move to Gloucester, Mary commented:
>
>> Unlike the people who have spent their whole lives here, we have travelled the world living in many different places. We chose Gloucester to live in. We are not here by circumstance; we have picked this place as the community that we want to live in. So we are keen to see the town go well. We saw the potential when we moved here ....We have become involved in many different groups and I guess I have taken on a leadership role in some of the groups.
>
> *Leader 2:*
> Chris moved to Gloucester in the mid-1990s. Chris moved to the region due to lifestyle reasons. An outdoor enthusiast, Chris quickly recognised the tourism potential of the region's landscapes. Chris was 'middle-aged' when he moved to Gloucester, and had spent many years living in Europe. Chris had been involved in numerous community-building activities in European towns, and therefore had gained a great deal of experience in leading endogenous development efforts. Unlike Mary, Chris did not view Gloucester as his 'final destination', rather he only intended on living in the area for as long as he felt it suited his lifestyle ambitions. Chris did leave the community after the first SnowFest.
>
> *Names changed to protect privacy

leaders' objectives. The leaders also faced opposition from some residents as they were 'new' to town and were seen as 'outsiders' wanting to change the town. The leaders attributed this resistance to some long-term residents perceiving the town as a service centre to primary industries situated in the hinterland. The leaders promoted the view that the town needed to shift away from this role to one where the town itself was a site for generating (rather than just servicing) core economic activities. Some residents believed that the leaders' push to expand the tourism industry in Gloucester threatened to decouple the town from the hinterland and thereby threatened the already fragile identity of the town.

Following the lack of support from residents and businesses for SnowFest, the leaders recognised that they needed to redesign their proposal to address community concerns. The leaders understood that wide community support for the festival was essential for realising the objectives of the project and that this was best achieved by working with the community to guide change rather than impose change. They therefore repositioned their proposal so that tourism was being promoted as a secondary or complementary industry for the town and was not threatening the agricultural or forestry sectors.

Community resistance to the project was partly a result of opposition to the individual leaders. As they had been living in the community for less than 10 years, they were seen by quite a few residents as 'outsiders', and as they did not initially work with established community groups or within established social networks they were seen as being in opposition to those groups and networks. The leaders, therefore, faced great difficulty in attracting volunteers to participate in the project. To overcome this they approached a number of core individuals, each of whom was well known and respected in the community, to take on a transactional leadership position. These individuals were given the task of recruiting other volunteers and ensuring all the activities necessary for making the event run efficiently were completed.

The volunteer 'transactional leaders' coordinated tasks such as organising a venue for the event, organising security for the event, arranging the trucking of the snow, securing sponsors for the festival and associated activities (including food and craft stalls) and organising the clean-up for the event. The leaders also negotiated support from the Gloucester Shire Council, which was critical in catalysing wider public support as many of the individual councillors also held influential roles within other community or sporting organisations. To gain the support of the individual councillors, the leaders invested considerable time into providing detailed information about their vision for the community and why the SnowFest festival was an appropriate mechanism for achieving this vision.

Commenting on the leadership demonstrated by the two leaders, a well-known local resident explained that:

> Those moving into Gloucester have often come from the city and from high powered jobs in management. Lots of them are semi-retirees and have had successful careers running businesses or working in leadership roles. They bring this drive and skills to the town. Many of the new ideas have come from these people. They have the knowledge and skills to pull off some crazy ideas such as the SnowFest. Many people thought that SnowFest couldn't work, but a handful of people managed to get it up and going and the town turned out on the day. So leadership at the moment is also [in addition to the Gloucester Shire Council] coming from these new people to town.

The first SnowFest festival, in July 2000, was widely regarded as a success. The organisers had prepared for approximately 2000 visitors;

Local Leadership and Rural Renewal through Festival Fun 69

however, more than 8000 visitors attended. One of the festival's leaders commented:

> The idea of SnowFest was to create a tourism event that attract[ed] spending into the town and showcase[d] the area. For every $5 visitors spent they would receive a 'snow dollar' which could be used at an auction at the end of the event. Auction prizes included a trip to Zurich. Spending topped $110,000.

The organisers of SnowFest achieved the leaders' goal of creating a tourism opportunity that attracted people to the town. SnowFest encouraged visitor spending in the town, and stimulated local businesses to shift their activities to incorporate facilities and services for the tourism market. The value of SnowFest to the community was even discussed in the Australian Senate, the Upper House of Parliament, with Senator John Tierney commenting:

> A more sustainable idea [for economic development] is the Gloucester snow festival in July. This occurs every year and could do with some additional financial support to make this a centrepiece of tourism in Gloucester. ... It can, through the SnowFest, generate greater levels of economic development. Last year, the attraction to the town was over 8000 people, so it was an event that doubled Gloucester's population for the weekend. (Tierney, 2001: 23137)

In 2001, the New South Wales government provided funding to employ a part-time event administrator, through the Townlife Development Program, after the two leaders prepared a successful bid for funding. In 2002, the event administrator became the full-time event organiser. This allowed the two leaders to step down from their active leadership duties and pursue more passive mentoring and support roles. In 2003, SnowFest had become so popular that over 12,000 people attended. More than 110 stores sold local produce, products and services over the weekend. However, in 2004, the event lost the financial support of the Townlife Development Program and one of the key original leaders withdrew from the organising team following the unexpected death of her husband. The organising team in conjunction with the Gloucester Chamber of Commerce decided to cancel the festival claiming that without funding from the Townlife Development Program the event could not afford a coordinator. The organisers felt that the event had grown so large that it could not be organised only with volunteer effort. ABC news reported:

> The Gloucester Chamber of Commerce cancelled this year's SnowFest, citing a massive budget increase which the community

was unable to sustain. President Peter Markey says instead a 'chillout day' is being held at the end of the month. He says the community recognises the value of the landmark festival, but SnowFest has become too large and difficult to run (ABC News, 5 July 2004).

Although SnowFest was ultimately a short-lived festival, during the years it operated, it catalysed community and business investment in the tourism industry. Prior to SnowFest, tourism was viewed as being a marginal economic activity for Gloucester, but the festivals showcased to the community the value of tourism. They also highlighted the role that local residents could play in shaping the economic and social adjustment processes that were occurring in the town. The nature of leadership in this project was particularly important to its success. The leaders displayed transformational leadership, through which they mentored others to undertake the necessary transactional leadership roles.

In 2008, the Gloucester Shire Council and Gloucester Chamber of Commerce made the decision to provide a small amount of financial and in-kind support for the festival. A small group of local residents, many of whom had been involved in the previous festivals, volunteered to organise and run the festival once more. The 2008 festival was a 'scaled-down' event compared to the large 2003 festival. Nevertheless, community members supported the event, turning out on the day in large numbers to participate in the various activities. The 2008 event saw the organisers return to 'grassroots' activities involving local businesses and community groups. The 2008 festival was not a centrepiece for Gloucester's tourism activities that year. Rather, it was a small event which complimented numerous other festivals and activities being held in the region.

In 2009, Gloucester hosted a number of different festivals, including the Shakespeare on Avon festival, the Mountain Man Tri Challenge and the Gloucester Spring Festival. A large number of new tourism-oriented businesses have been established and the Council has developed a mosaic of parks and recreational spaces for visitors. Many of the individuals who were involved in the SnowFest festival now have influential roles in organising these other tourism and community events.

Leadership and the Broader Value of Festivals

Transformational leaders exhibit behaviour that make followers aware of the importance of their involvement in tasks, activate followers' higher-order needs and encourage them to move beyond self-interests for the sake of the wider community (Podsakoff *et al.*, 1990). On the other hand, transactional leaders exhibit behaviours 'founded on an exchange process in which the leader provides rewards in return for the subordinate's efforts' (Podsakoff *et al.*, 1990: 108). In the SnowFest project, the nature of the project's objectives necessitated both transactional and transformational leadership. Transactional leadership was important in day-to-day organisational task such as organising the venue and securing sponsors. Transformational leadership was critical in attracting volunteer participation and motivating community support for the event.

Transformational leadership is founded on three behavioural factors. The factors are summarised as 'charisma/inspirational', 'intellectual stimulation' and 'individualised consideration'. In the Gloucester SnowFest case, the leaders displayed each of these factors throughout the project. In terms of the 'charisma/inspirational' factor, they purposefully engaged with established community leaders to translate to the wider community a clear vision for the project and a justification for how it aligned with the existing community values. Through this strategy, the leaders provided the opportunity for members of the wider community to identify with the vision. To further engage community support, they displayed the depth and breadth of their knowledge on community 'revitalisation' efforts through directly talking to potential supporters and volunteers. This activity involved a lot of 'leg work' and also opened them up to criticism. However, this direct approach enabled potential supporters and volunteers to gain confidence and trust in the leaders' abilities. Furthermore, they put particular effort into promoting the project as one that would deliver tangible benefits to local businesses and community groups. This demonstrated their willingness to work for a higher-order interest, which in turn encouraged followers to adopt a similar behaviour.

In terms of 'intellectual stimulation', the leaders encouraged followers to take ownership of various parts of the project. Individuals were encouraged to contribute ideas that they thought might enable the project to achieve its overall objective. In terms of 'individualised consideration', the leaders worked largely behind the scene to ensure that all volunteers were satisfied with their engagement with the project. Where possible, they provided volunteers with training, advice and motivation.

It is inherently difficult to quantify the effect of transformational leadership on the community or indeed to the role of the festival in stimulating ongoing leadership in the community. One possible way of assessing the impact of transformational leadership is through reviewing the impact that SnowFest had on the community. Often measured in terms of visitor numbers and expenditure or investment in new tourism enterprises, it has been widely agreed that SnowFest made a considerable contribution towards catalysing community social and economic investment in Gloucester's tourism industry. SnowFest increased community awareness of the potential social and economic contribution of the tourism sector. SnowFest also demonstrated the importance of community involvement in generating new economic activities. As a result of being involved in SnowFest and seeing the potential for growth in Gloucester's tourism sector, a number of individuals established businesses to service the tourism sector. These businesses have been successful in catalysing further activity and investment in the tourism sector. Local businesses, organisations and the Council also increased their support for emerging tourism activities and businesses. In this case, had the leaders only adopted a transactional leadership style they would not have achieved successful outcomes.

While considering the value of transformational leadership by examining the endpoints of a project is useful, it is also important to consider that transformational leadership resulted in attitudinal and value change during the project. The leader's sharing of knowledge about community revitalisation and the role of local initiatives in a tourism-led recovery was particularly important in catalysing an attitudinal shift in how tourism was viewed in the town. Prior to SnowFest, tourism was viewed either as being in competition with the primary industries or as being secondary to these traditional industries. Through the organisational process, and the particular approach taken by the leaders, these attitudes were reversed. There was also a shift in perception regarding the value of local participation in influencing socio-economic adaptation processes. Festivals are important sites of leadership formation and development, while transformational leadership facilitates community-led socio-economic adaptation efforts. This, in turn, provides support for ongoing government and philanthropic efforts to develop transformational leadership capacities in rural communities.

References

Australian Broadcasting News (ABC News) (5 July 2004) Hopes melt away for Snowfest. ABC News.
Avolio, B.J. and Bass, B.M. (1999) Re-examining the components of transformational and transactional leadership using the Multifactor Leadership Questionnaire. *Journal of Occupational and Organisational Psychology* 72, 441–462.
Bass, B.M. (1985) *Leadership and Performance beyond Expectations*. New York: Free Press.
Bass, B.M. (1988) The inspirational process of leadership. *Journal of Management Development* 7, 21–31.

Bass, B.M. (1999) Two decades of research and development in transformational leadership. *European Journal of Work and Organizational Psychology* 8, 9–32.

Bass, B.M. and Avolio, B.J. (1990) *Transformational Leadership Development: Manual for the Multifactor Leadership Questionnaire.* Palo Alto, CA: Consulting Psychologist Press.

Bass, B.M. and Avolio, B.J. (1993) Transformational leadership: A response to critiques. In M.M. Chemmers and R. Ayman (eds) *Leadership Theory and Research: Perspectives and Directions* (pp. 49–88). San Diego, CA: Academic Press.

Bass, B.M. and Avolio, B.J. (1994) *Improving Organisational Effectiveness through Transformational Leadership.* Thousand Oaks, CA: Sage.

Burns, J.M. (1978) *Leadership.* New York: Harper & Row.

Davies, A. (2007) Organic or orchestrated: The nature of leadership in rural Australia. *Rural Society* 17, 139–154.

Davies, A. (2009) Understanding local leadership in building the capacity of rural communities in Australia. *Geographical Research* 47, 380–389.

Egan, R.F.C., Sarros, J.C. and Santora, J.C. (1995) Putting transactional and transformational leadership into practice. *The Journal of Leadership Studies* 2, 100–123.

Haslam-McKenzie, F. (2002) Leadership development: Flogging a dead horse or the kiss of life for regional Western Australia? *Sustaining Regions* 1, 24–31.

Kotter, J. (1998) Cultures and coalitions. In R. Gibson (ed.) *Rethinking the Future: Rethinking Business, Principles, Competition, Control and Complexity, Leadership, Markets and the World* (pp. 164–178). London: Nicholas Brealey.

Parry, K.W. (1999) The new leader: A synthesis of leadership research in Australia and New Zealand. *Organizational Studies* 5, 82–105.

Podsakoff, P.M., MacKenzie, S.B., Moorman, R.H. and Fetter, R. (1990) Transformational leader behaviors and their effects on followers' trust in leader, satisfaction, and organisational citizenship behaviors. *Leadership Quarterly* 1, 107–142.

Sarros, J.C., Cooper, B.K. and Santora, J.C. (2008) Building a climate for innovation through transformational leadership and organisational culture. *Journal of Leadership and Organisational Studies* 15, 144–158.

Shamir, B., House, R.J. and Arthur, M.B. (1993) The motivational effects of charismatic leadership: A self-concept based theory. *Organisational Science* 4, 577–594.

Tierney, Sen. J. (2001) *Gloucester, New South Wales: Job Losses.* Speech. Tuesday, 27th March, 2001 (pp. 23135–23137). Canberra: Commonwealth of Australia Parliamentary Debates.

Chapter 5
Economic Benefits of Rural Festivals and Questions of Geographical Scale: The Rusty Gromfest Surf Carnival

P. TINDALL

Introduction

Many of Australia's non-metropolitan areas face an uncertain future (Murphy & Murphy, 2001), hence frequently turning to new, non-traditional means of income, such as tourism, to sustain them (Hall, 1992; Murphy & Murphy, 2001). Carnivals, festivals and sports events are some of the many ways that vulnerable rural communities seek to generate economic activity. This chapter investigates the financial benefits which special events can bring to rural towns and discusses the methodologies employed in economic impact studies. A comparison of the results of various economic impact studies is undertaken to examine the influence of the size of the festival and its host town on relative economic impacts. Through a comparison of one case study – a surfing carnival at the small Australian town of Lennox Head – with other economic impact studies of a variety of larger, metropolitan events, I explore how economic benefits must be understood as a function of geographical scale.

Economic Impacts of Special Events

The cumulative economic impact of festivals has the potential to be significant to rural economies. Moreover, not all festivals demand costly infrastructure. Surf competitions are one example of festivals that often do not require the large capital costs associated with construction works, such as athletics tracks, sports stadiums or other costly amenities. Sports festivals can therefore be a useful source of income for small communities, without significant local investment. Smaller sporting events may have a substantial cumulative impact on the economy of a community. This impact may be as, if not more, significant in a relative sense than one-off large-scale events (Hall, 1992), and have a more significant impact on smaller towns than larger ones, with a greater per capita flow of income into the community.

Many studies have attempted to measure the impacts of special events upon communities, although the majority have been concerned with large events, even mega-events (also known as hallmark events), with few concerning themselves with smaller festivals (Hall, 1992; Taylor, 2003). While economic studies have been carried out in towns and cities of various sizes, few researchers have taken the opportunity to analyse, or even note in passing, that the size of the town may be important when determining the amount of economic influence the event will have on the community.

Economic impact studies are usually carried out in order to justify the government or organiser's view on the economic sustainability of a given event, and to justify the reduction, maintenance or increase in funding for that event (Crompton *et al.*, 2001; Davidson & Schaffer, 1980). They aim to measure how much the event expands the economic base of the community (Hone, 2005). Studies are rarely carried out simply to determine the public costs and benefits of an event on behalf of the community itself.

What makes the assessment of economic impacts more difficult is that festivals vary greatly in nature, covering a wide variety of host locations, time frames and audiences. Some are held over a single day in one specific place with easily estimated attendance levels, while others are a mass of various activities, over a wide area lasting for a number of days or weeks, at which attendance is not so easily estimated (Davidson & Schaffer, 1980). Even the simple geographic location of the central area of festival activities in relation to the town's main street matters: in Parkes, the Elvis Revival Festival (Chapter 11) boomed once it moved from a city fringe park to the central town square; whereas at Bundanoon local shops lost business when the Brigadoon Scottish Festival outgrew the main street and had to move to large showground premises 500 m away (Chapter 16). At the Deniliquin Ute Muster, enormous revenues are generated, but the spatial separation of the festival from the town (necessary because of the size of the event, and associated noise and traffic impacts) reduces benefits flowing to main street businesses.

Not surprisingly, a number of different methodologies have been put forward to estimate the economic impacts of festivals. Determining the most appropriate method for any given study is crucial, generally dependent on time and funding constraints, and knowledge of the scale and nature of the event. Conducting an economic impact assessment of a special event involves the collection of initial data relevant to various types of expenditure associated with it. This includes expenditure by event organisers and sponsors, capital expenditure and opportunity costs for facilities related to the event as well as the expenditure generated by visitors to the event (Allen et al., 2002).

The main methodological difference involved in the initial gathering of data to form the basis of an economic impact assessment, is whether to study the supply side of expenditure (i.e. the business owners), the demand side (i.e. the visitors) or a combination of the two (Burgan & Mules, 1992; Davidson & Schaffer, 1980; Gelan, 2003). Most studies have concentrated on assessing economic impacts by studying the demand side of the equation, using visitor surveys, interviews and expenditure diaries (Burgan & Mules, 1992). But methodologies can be problematic, and there is a need to tailor them to suit individual events. The most common type of visitor surveys are 'on-the-spot' surveys (Davidson & Schaffer, 1980; Gelan, 2003). These surveys involve either intercepting spectators on their way into or out of the event at set locations, or using roaming interviewers to survey people throughout the event site. The diary method of surveying visitors involves recruiting respondents before their visit has commenced, and having them keep a diary of all their expenses which is mailed back to the surveyor (Davidson & Schaffer, 1980).

A comparison of on-the-spot surveys and the diary method undertaken by Faulkner and Raybould (1995) found that expenditure diaries were more accurate. However, they have a substantially lower return rate and demand more time for diaries to be returned to researchers. As respondents are asked to complete on-the-spot surveys before both the event and their visit have ended, it is difficult to gain an accurate account of how much they will spend. Tourists have been known to consistently underestimate their spending at short-term events (Irwin et al., 1996). It can also be difficult for visitors to differentiate between event and non-event-related expenditure, especially for those whose visits are not entirely event related (Allen et al., 2002; Davidson & Schaffer, 1980; Howard & Crompton, 1995). Open-entry events without set entry and exit points (such as surfing carnivals) can make undertaking visitor surveys difficult, due to inherent random sampling, sample size and bias issues (Crompton, 1999; Department of Sport, Recreation & Tourism, 1986).

The use of business surveys to estimate the impact of the event on business turnover is far less common. One of the problems with this methodology is the reluctance of many private business owners to divulge

Economic Benefits of Rural Festivals 77

event-specific turnover figures for privacy reasons (Davidson & Schaffer, 1980). Curiously, there have not been suggestions forthcoming in the literature as to how to best avoid this issue. One solution, tested in this study, may be to ask about changes in turnover and derive monetary impacts by calculating the percentage change in relation to the normal turnover. Overall, businesses were found to be more responsive to this approach.

It can also be difficult for businesses to determine how much revenue is attributable to an event (Burgan & Mules, 1992; Department of Sport, Recreation and Tourism, 1986; Faulkner & Raybould, 1995; Gelan, 2003). It is generally believed that the majority of business owners are not able to determine which customers are in the area specifically for the event and those that are in town for other reasons. However, as the majority of studies to date have been carried out in large cities, where it is indeed difficult to differentiate between regular customers and visitors, this is not surprising. In a rural village or small town however, it is more likely that business owners are able to better estimate the amount of income attributable to the event, due to the close vendor–customer relationships inherent in small communities.

It can also be difficult to determine the amount of visitor expenditure that would have occurred regardless of the event. Only 'new' expenditure attributable to the event should be included in calculations (Crompton *et al.*, 2001; Gelan, 2003; Hall, 1992). Viewpoint is an important issue here, as it is necessary to decide the geographical boundaries of the study area to determine who is a 'visitor' and who is a 'local'. As Burns and Mules (1989) pointed out, the smaller the area of focus, the larger the number of people who fall into the 'visitor' class and therefore the greater the expenditure considered 'injected'. On the other hand, the leakages, or money lost to the town through the purchasing of goods from other areas, will also be greater (Allen *et al.*, 2002; Burns & Mules, 1989; Department of Sport, Recreation and Tourism, 1986; Hall, 1992).

In this chapter, I compare one festival from my own research – the Rusty Gromfest surf carnival – with a number of other events (studied by others in larger, metropolitan areas), in order to demonstrate variations in interpreting economic impact. I focus especially on geographical scale, an issue particularly relevant for rural festivals.

The Rusty Gromfest

The Rusty Gromfest, hereafter referred to simply as Gromfest, is a youth surf event held over a period of four days in the town of Lennox Head, a small town on the far north coast of New South Wales. Every year the event attracts over 300 competitors (Rusty Gromfest, 2005). The event has become the largest and most widely recognised event in Australian youth surfing (*The Lennox Wave*, 2004). It attracts competitors from across Australia as well as a substantial contingent of international competitors. Event organisers estimated crowd numbers for the 2005 event (at which this study was conducted) to be around 5000 (Max Perrot, Event Director, Rusty Gromfest, pers. comm., 2005). This figure is inclusive of repeated attendances on subsequent days, and is an estimate of the total number of people who would have 'passed though the turnstiles' (had there been any), and includes visitors, local residents, media, sponsors, surf industry representatives and judges (see below for a more precise estimation of daily attendance figures).

Gromfest usually begins on a Friday morning and runs over the entire weekend until the following Monday afternoon. In 2005, the event was delayed by a day due to severe flooding in the surrounding areas. The event began on Saturday morning and continued through to Tuesday. This meant there were a number of empty rooms in accommodation establishments, as people could not get to town due to flooding, and others delayed their journey for a day to arrive on the Saturday. This negative impact was alleviated by the extension of other visitors' trips, with some remaining until the Tuesday.

Lennox Head (population 2514) is 10 km north of the large township of Ballina (population 20,000), and 20 km from Byron Bay (population 10,000). Due to its proximity to Ballina and Byron Bay, Lennox Head has the unusual ability of being able to host an event with the potential of attracting more visitors than the town can accommodate, with Ballina and Byron Bay able to absorb the overflow. The proximity of a domestic airport in nearby Ballina, which provides direct flights to Sydney with all major domestic carriers, means that Lennox Head is easily accessible to both domestic and international visitors.

Methods

A mixed methods approach was used, combining surveys of visitors and local businesses. Visitor surveys were used to determine the amount of expenditure generated in Lennox Head. Respondents were asked how many days they planned to attend the event, and how much money they intended to spend in various categories, differentiating between expenditure within Lennox Head and elsewhere on the trip. The business survey aimed mainly at collecting information relating to the proportion of extra turnover received as a result of the event. In lieu of multiplier or cost–benefit analysis, businesses were asked how much of their stock was imported from outside the town, in an attempt to quantify the amount of direct expenditure that remained within Lennox Head.

Burns and Mules' (1989) study of the Adelaide Grand Prix was considered groundbreaking in assessing the economic impacts of special events in Australia (Allen *et al.*, 2002; Gratton *et al.*, 2000; Hall, 1992) for their use of a combination of economic multipliers and cost benefit analysis methodologies. Burns and Mules calculated two sets of total economic impact results, giving both an 'upper' and 'lower' bound result to account for estimated maximum and minimum spending costs. A similar format, with the addition of a middle bound, was utilised to allow for errors in responses pertaining to expenditure, while also allowing for those casual visitors and time switchers (those who would visit Lennox Head anyway at some other time, but 'switched' to attend while the specific event was taking place) who were wrongfully identified through misinterpretation of questions, and to also take into account the suggestion of Hone (2005) that some of their expenditure should indeed be included in calculations. A more detailed description of this methodology is included in Tindall (2005).

Who responded to the visitor survey?

One hundred and eight completed visitor surveys were received. The total number of days spent attending Gromfest in Lennox Head by these 108 groups was 345. In total, this covered the spending for 421 people including children, and so by multiplying the number of days of attendance by the total number of people for each individual group, we come to the conclusion that the total number of visitor days spent attending Gromfest was 1360. The average number of people per visitor group was 3.8.

Fifty-five per cent of visitors were under the age of 25. Some 34 visitors (8.4%) were aged between 26 and 40 years, while 137 visitors (33.7%) were aged between 41 and 60). Twelve visitors (2.95%) were aged over 60 years. The average age was at the upper end of the 11–25 years group, with data bimodal around the 1–25 and 41–60 years categories, indicating the strong presence of family groups containing parents and children. Indeed, the most common visitor group types were those containing 1 adult and 1 child (17%), 2 adults and 2 children (14%), 1 adult and 2 children (13%) and 2 adults and 1 child (12%). The number of groups that did not have any children was low, with only three sampled single adults and four childless couples attending the event.

The highest proportion of visitors came from New South Wales (68.5%), from the Sydney, Newcastle and Wollongong region (29 surveyed groups or 27%), followed by the region surrounding Lennox Head (25 respondents, or 23%). Twenty-four groups (22%) came from Queensland, and four were from other areas of Australia, including one from South Australia and three from Victoria (0.9% and 2.8% respectively). Six surveyed visitor groups (5.5%) travelled to Lennox Head from overseas to attend Gromfest, including four from New Zealand (3.7%) and one each from England and the United States (0.9% each).

Just over half of the visitor groups stayed within the town of Lennox Head itself (54%). A further 22% of surveyed visitors stayed in the nearby towns of Ballina, Byron Bay and Suffolk Park. A small proportion stayed within the Tweed Coast, Coolangatta and Gold Coast regions, located further to the north. Seven percent of respondents stayed 'at home', indicating that they were day visitors.

The high percentage of visitors not staying in Lennox Head (46%), while having a positive impact on surrounding areas, could also be seen as a missed opportunity for Lennox Head, by drawing visitors away from the town and providing opportunities for visitors to spend money elsewhere. On the other hand, the close proximity of these other localities enabled a greater overall attendance at Gromfest, as they are capable of accommodating additional visitors above what accommodation in Lennox Head can support. The geographical proximity of neighbouring towns matters when interpreting economic impact, though exactly how benefits are distributed is far from being clear cut.

Who responded to the business survey?

Sixty-five business surveys were returned completed, a return rate of 77.4%. The survey sample included businesses from a wide variety of industries, with a number of services being represented. It is safe to assume that because surveys were offered to all identifiable businesses within Lennox Head, there was a very representative sample. The high response rate of 77.4% supports this claim. Nineteen food service businesses returned surveys, making this the most highly represented industry group (29.2%), followed by service industry businesses (nine businesses; 13.9%), retail (eight businesses; 12.3%) and accommodation establishments (seven businesses; 10.8%). Five each of real-estate agents and health and beauty establishments responded, as did four clothing shops, three grocery stores and both service stations within the town. All the surveyed businesses employed less than 20 people.

Calculating the Direct Economic Impacts of the Gromfest Surf Carnival

Visitor expenditure

Three sets of calculations were carried out from visitor expenditure estimates in the survey: lower-, middle- and upper-bound calculations. The lower-bound calculations were based on data from the 40 respondents who were not identified as casuals or time switchers. The upper bound was calculated using the expenditures of all 108 visitors who responded to the question, while the middle bound included the 40 previously identified as not being casuals or time-switchers, as well as the expenditure of half the remaining visitors (74 in total). This middle-bound estimation allowed for some respondents experiencing confusion over the time-switcher and casual questions, and for errors in the spending estimates. The expenditures for the visitors in each calculation were added together to generate the total amount of expenditure.

An attempt was made to estimate the total number of visitors who attended the event, and to extrapolate the actual spending results to give an indication of the real impact of the event. There were 476 competitors registered for the event (Rusty Gromfest, 2005). It was estimated that two-thirds of these would be non-local competitors (317 competitors). This figure does not allow for there being more than one competitor within each visitor group. However, it does allow for non-competing spectator groups to be counted. From the survey, it was known that there was an average of 3.7 people in each visitor group, which for the purpose of this exercise was rounded up to 4. The average competitor would then have 4 people in his or her immediate travelling group, including themselves, making a total of 1268 visitors ($4 \times 317 = 1268$) – a figure very similar to that provided by organiser, Max Perrot (pers. comm., 2005). It was also known that the visitor expenditure data collected in the field were for 421 people. In order to extrapolate the data to 1268 people, the actual expenditure data collected in the field was multiplied by 3.015 ($1.268/420.5 = 3.015$). The result is the total amount spent in Lennox Head by an estimated 1268 visitors for their entire stay in the town.

In order to make these figures more meaningful, they were converted into a similar time frame as the business data. Businesses were asked to record their turnover for the 'weekend' of the event, that is Saturday and Sunday, and did not include turnover received on Friday, Monday or Tuesday. Visitors were asked to record their expenditure for their whole trip, which could vary from 1 day to 3 weeks. It is common practice within the literature to divide the amount of expenditure for the whole trip by the

Table 5.1 Total spending by all surveyed visitors and average per visitor group in Lennox Head on the weekend of Gromfest

	Total for industry group			Average per visitor group		
Industry group	Lower	Middle	Upper	Lower	Middle	Upper
Accommodation	$9,093	$24,964	$41,149	$227	$342	$388
Food & beverages	$7,380	$17,570	$26,794	$184	$241	$253
Retail shopping	$2,828	$7,773	$11,499	$71	$106	$108
Private auto expenses	$1,815	$4,384	$6,072	$45	$60	$57
Night clubs, bars, pubs, etc.	$892	$1,676	$3,208	$22	$23	$30
Other	$0	$3,111	$3,570	$0	$42	$33
Rental car expenses	$0	$0	$0	$0	$0	$0
Total	$22,008	$59,479	$92,292	$323	$472	$482

number of days the event has occurred to give a per-day expenditure. Following this, the visitor expenditure was then divided by 5 days to get a per-day expenditure and multiplied by 2 to give their expenditure for only Saturday and Sunday. This then provided the 'weekend' expenditure of the visitors to Gromfest.

Those visitors to Lennox who participated in the study spent a total of between $22,008 (lower-bound estimation), $59,479 (middle-bound estimation) and $92,292 (upper-bound estimation) (Table 5.1). The largest amount of money was spent on accommodation ($9093 lower bound; $24,969 middle bound; $41,149 upper bound), followed by money spent on food and beverages ($7380 lower bound, $17,570 middle bound; $26,794 upper bound), and retail spending ($2828 lower bound, $7773 middle bound and $11,499 upper bound). Spending in night clubs, bars and pubs was relatively low due to the peculiar demography of visitors (families with children).

Business turnover

The initial step in calculating the amount of business activity attributable to Gromfest was to determine for each industry group what the average turnover was for a normal weekend during the June/July period, and then the average turnover change for all businesses on the weekend of Gromfest. These averages, calculated for each industry group, were then applied to those within each group who did not respond to one or both of the necessary survey questions. Reponses for those who answered the questions were not altered; only those with one or the other missing were calculated. Three sets of calculations were carried out, using an upper, middle and lower bound of both percentage change and turnover. These figures were then extrapolated to include all 84 identifiable businesses within the town. As 77.4% of businesses responded to the survey, the results were divided by 77.4 then multiplied by 100 to cover all businesses in the town. This allowed for a more meaningful comparison with the visitor expenditure results.

In order to determine the amount of activity attributable to Gromfest, the average turnover on a normal weekend in June/July was subtracted from the amount of turnover received on Gromfest weekend. The remainder was the amount of turnover that was attributed to Gromfest by business respondents, that is the extra turnover they received as a result of the event.

The total amount of turnover attributed to Gromfest was $31,432 (lower bound), $66,251 (middle bound) and $109,214 (upper bound). Estimates of turnover attributable to Gromfest varied greatly between industry groups. Real-estate business attributed the highest amount of expenditure to Gromfest, with estimates of between $16,795 (lower bound) and $44,574 (upper bound), followed by food shops, with estimates between $9264 and $39,124. All remaining industry groups attributed less than $9000 to Gromfest in all three bounds, showing that they experienced a modest economic impact. Health and Beauty and Service businesses consistently received the smallest amount of turnover as a result of Gromfest, with less that $400 in extra turnover per business attributed to the event. The service industry attributed none of its turnover on Gromfest weekend to Gromfest. As these estimates are directly dependent on those of the average normal turnover and turnover on Gromfest weekend, it is also possible that accommodation has been underestimated, and that any underestimation would be compounded through extrapolation.

The total turnovers attributable to Gromfest figures for all industry groups were worked out as a percentage of the corresponding total turnover for all industry groups on a normal weekend in June/July, giving the percentage of normal turnover that was received on Gromfest weekend. It was found that Gromfest brought between 19% and 44% of extra turnover to businesses in Lennox Head (lower and upper bounds respectively), with a middle-bound estimation of 32%.

Reflecting these findings, the 2005 Gromfest did not result in the direct creation of any new jobs, either full time or part time. However, Gromfest was found to have generated at least 38 extra shifts within the surveyed businesses, which in itself is an indication that additional money was brought to the town. Two-thirds of the new shifts generated by Gromfest were in food establishments. One accommodation establishment put on four extra staff, while one grocery business put on two extra staff for the weekend. Two retail clothing stores put on one or two extra staff.

In the survey, 25 businesses (42%) indicated that less than 25% of customers were tourists. Eighteen businesses (31%) indicated 25–50%; nine businesses (15%) indicated 50–75% and seven (12%) indicated over 75% of customers were tourists. Predictably, those most likely to say that over 75% of customers were tourists were accommodation providers and food shops. Least likely were services, petrol stations and health and beauty establishments. Only 10 of the businesses surveyed (15%) were approached to supply goods and services for the Gromfest carnival, probably reflecting the highly specialised nature of the event.

Event Comparison: A Question of Scale

How does one interpret such results on economic impact? At one level, the raw numbers suggest that contributions to the local economy are marginal: the total dollars earned do not seem particularly impressive; few extra people were employed, and although noting increases in turnover consistently, businesses only gained modestly on average. However, what needs to be discussed in addition to calculations of direct economic impact are issues of the geographical scale of the place hosting an event, and the potential for the staging of many festivals in such places. As Gibson *et al.* (2010) demonstrated, festivals may be significant to small rural communities because of their sheer proliferation – even if singularly insignificant – in places with small populations (and thus the potential for high per capita circulation of benefits).

One way of demonstrating this is by comparing results from the Gromfest surf carnival in Lennox Head with four other Australian events where comparable methods were deployed. Such comparisons are useful, because they demonstrate how per capita impacts vary depending on the size of the host community, and the size of the event. In case of the events compared here, they are much larger than Gromfest, and are hosted in places of varying size. From this comparison, we can see that scale is indeed an important factor in determining the relative impact of a special event. The four other events compared here are the Australian Open Tennis

Tournament (held in Melbourne, a city of over 3 million people, and subject to economic impact analysis in Downey, 1991); the World Cup of Athletics (held in Canberra, with a population of 270,000 at the time of the event when analysed by the Australian Department of Sport, Recreation and Tourism, 1986); the National University Games competition, held in the NSW country town of Lismore (with a population of 43,000 at the time studied by Walo et al., 1996); and another surf carnival, held in Torquay, Victoria, a town of nearly 3000 residents (and subject to analysis in Downey, 1991). There are some methodological variations across these studies, which in the interests of brevity are not discussed in detail here (see Tindall, 2005). Notwithstanding methodological variations, clear insights can be gained on the *relative* economic impacts of events, an issue of particular importance for rural communities.

Event duration is an important factor to consider when comparing the economic impacts of different events (Hall, 1992). The Ford Australia Tennis Open was held over a two-week period, and so it is not surprising for it to have had the greatest total economic impact (Downey, 1991). The Ford Australia Tennis Open brought $29.9 million to its host community, while the shorter events brought substantially less (Table 5.2). Gromfest, being a short event, also resulted in substantially less total economic impact (cf. Gratton et al., 2000). With the exception of the 1985 World Cup of Athletics, per visitor per day spending tends to increase with duration.

Explanations of why per visitor per day spending was substantially lower for Gromfest than for the other four events may include the small proportion of visitors who stayed within the town and the lack of opportunities to spend money while at the event (Tindall, 2005). Attendees at the surf carnival were not wealthy (in comparison to say, fans at the Australian Open tennis tournament) and opportunities to spend were rarer, with few souvenirs available and only a small number of local businesses open to go shopping for unplanned, discretionary purchases. With only 54% of visitors staying within Lennox Head itself, the average spending on accommodation within the town was low, as was spending on food and beverages. There were few opportunities for spectators and competitors to spend money at Gromfest, with only two stalls on site (one event sponsor selling surfboard fins and the the other a sausage sizzle). Spectators needed to leave the event site (the beach) and walk to the shops and cafés in the main street of town for their purchases. Surf events are unusual in this respect. At larger sports competitions, especially in big cities, money-spending opportunities are more common (souvenirs in stadiums, shops in nearby inner-city districts), generating higher per visitor per day spending.

Table 5.2 Comparison of the direct economic impacts of various Australian sports events

Event	Location	Period of economic impact	Direct economic impact (1) ($1000)	CPI (inflation) multiplier	Direct economic impact (2) ($1000)	Visitor attendance	Per visitor per day expenditure	Population at time of event[a]	Impact per capita of host population	Per day impact per capita of host population[b]	Source
1989 Ford Australia Tennis Open	Melbourne, VIC	14 days	$20,000	1.496	$29,920.00	312,000	$64	3,156,700	$6	$0.45	Downey (1991)
1985 World Cup of Athletics	Canberra, ACT	3 days	$12,100	2.578	$5,410.00	10,000	$71	270,000	$44	$15	Dept. Sport, Rec. & Tourism (1986)
1995 NCUSA University Games	Lismore, NSW	4 days	$393	1.253	$492.00	1,000	$58	42,954	$9	$2	Walo et al. (1996)

	Torquay, VIC	6 days	$1,822	1.496	$2,726.00	40,000	$62	2,892	$630	$105	Downey (1991)
1989 Easter Surf Carnival											
2005 Gromfest (lower bound)[c]	Lennox Head, NSW	4 days	—	—	$44.02	1,268	$9	2,514	$9	$4	Tindall (2005)
2005 Gromfest (middle bound)[c]	Lennox Head, NSW	4 days	—	—	$118.94	1,268	$23	2,514	$24	$12	Tindall (2005)
2005 Gromfest (upper bound)[c]	Lennox Head, NSW	4 days	—	—	$184.58	1,268	$36	2,514	$37	$18	Tindall (2005)

Note: All values were adjusted to 2005 values using the Consumer Price Index (CPI). Direct economic impact (1) refers to the direct economic impact ($AUD) at the time of the event. Direct economic Impact (2) refers to the direct economic impact ($AUD) adjusted for inflation to 2005.
[a] See Tindall (2005) for explanation of sources for population estimates.
[b] Per capita per day estimates were derived by dividing the total economic impact for each event by the size of the host community at the time of the event.
[c] Visitor spending estimates have been used.

The size of the event, as indicated by the number of attendees, is another determining factor in the economic benefits an event brings (Hall, 1992). Of the events compared in Table 5.2, it is not surprising to find that the events which had the highest visitor attendance numbers also generated the largest total economic impacts (see also Allen *et al.*, 2002: 38; and Daniels & Norman, 2003). There are notable exceptions to this, with the 1989 Easter Surf Carnival having a lower economic impact than the 1985 World Cup of Athletics, despite having 30,000 more attendees. Per visitor per day estimates were not affected by the size of the event, nor were per capita per day impacts; only total economic impacts varied. It is most likely that this was in turn an effect of the size of the host community, and not a reflection on the size of the event.

While the Australian Tennis Open was the largest of the five events and had the largest total impact, it only equated to an additional 45 cents per day for the host population when calculated per resident (Table 5.2). Canberra residents did much better from their hosting of the World Cup of Athletics, receiving an additional $15 per resident per day. Lismore, host community of the 1995 NCUSA University Games, received an additional $2 per resident per day. Considering this event attracted only 1000 spectators, this figure was relatively high, and was a result of the host population being considerably smaller than that of Melbourne. The size of the host community indeed appears influential in determining the *relative* scale of economic impact.

Figures from the other two events, Gromfest and the Easter Surf Carnival, support this claim, with Gromfest bringing between $4 and $18 per resident per day. Despite the event having the lowest per visitor per day expenditure and the smallest total economic impact, it had a *relative* economic impact on its host community greater than that of the Ford Australia Tennis Open and the NCUSA University Games. Upper-bound estimates for Gromfest were also greater than the World Cup of Athletics.

Further illustrating this sense of relative economic impact was the 1989 Easter Surf Carnival at Torquay, which attracted 40,000 spectators, over 30 times the size of the host population. The event brought $2.7 million dollars (adjusted for inflation) to Torquay, the second highest total economic impact of the five events. When divided by the small host population of 2892, this equates to an additional $105 per resident per day, demonstrating the impact that the hosting of a large-scale event can have on a small rural community. If a given event is held in different-sized towns, the smaller town would enjoy a greater per-business impact, simply because there would be less businesses between which to divide the total economic impact (Wall & Mitchell, 1989).

Conclusion

This chapter detailed an example of how to calculate economic benefits from a festival. Methodologies are far from universally accepted, but it is clear that there is some consistency in results between surveys administered to visitors and to businesses. In comparison, visitor expenditure estimates attributable to Gromfest were between $22,008 and $92,292, while business turnover estimates were between $31,432 and $109,214. Whereas, business surveys have been written off by some researchers as inaccurate, in small rural places where shopkeepers know their regular customers (indeed, may literally know just about everyone in town) identifying new business attributable to visiting festival attendees is possible. For this reason, those wishing to examine the economic benefits of rural festivals would do well to survey both visitors (in the conventional manner now often replicated in the field) *and* businesses. Both methods showed how economic benefits accrue to different segments of the local economy (food and accommodation more so than services or health and beauty establishments), and how the peculiar demography of visitors can affect resulting patterns of economic activity. Families with children participating in a surf carnival buy food and stay in hotels much as do all kinds of visitors to larger sporting spectacles, but their propensity to buy souvenirs, clothing, alcohol and concert tickets is markedly different, with resulting implications for how host places accrue benefits.

Much criticism has been levelled at mega-events, due to their high number of negative social and environmental impacts, and increasingly questionable claims about economic benefit (balanced against escalating infrastructure costs). The results of this research (and that of studies such as Gibson *et al.*, 2010) suggest that a change in policy focus is warranted. When economic benefits are understood in a relative sense, seemingly insignificant events in small places, with little or no need to build new local infrastructure, can be understood as economically meaningful. Given that such events have proliferated in recent decades (Gibson *et al.*, 2010), it is feasible to suggest that festivals (even if most remain small and community focused) can become, if they are not already, a substantial component of the local economy of small rural places. The hosting of smaller events is usually far less socially detrimental to the host communities than the hosting of larger events, and there is not the same level of demand on infrastructure and local facilities. The hosting of small-scale events which require minimal infrastructure or set-up costs is a worthwhile investment for small communities who do not have the infrastructure or finance for hosting larger-scale events.

References

Allen, J., O'Toole, W., McDonnell, I. and Harris, R. (2002) *Festival and Special Event Management* (2nd edn). Milton: John Wiley & Sons.

Burgan, B. and Mules, T. (1992) Economic impact of sporting events. *Annals of Tourism Research* 19, 700–710.

Burns, J.H. and Mules, T. (1989) An economic evaluation of the Adelaide Grand Prix. In G. Syme, B. Shaw, D. Fenton and W.S. Meuller (eds) *The Planning and Evaluation of Hallmark Events* (pp. 73–80). Aldershot: Ashgate.

Crompton, J. (1999) *Measuring the Economic Impact of Visitors to Sports Tournaments and Special Events*. Ashburn, VA: National Recreation and Park.

Crompton, J., Lee, S.K. and Shuster, T. (2001) A guide for undertaking economic impact studies: The Springfest example. *Journal of Travel Research* 40, 79–87.

Daniels, M.J. and Norman, W.C. (2003) Estimating the economic impacts of seven regular sport tourism events. *Journal of Sport Tourism* 8, 214–222.

Davidson, L. and Schaffer, W. (1980) A discussion of methods employed in analyzing the impact of short-term entertainment events. *Journal of Travel Research* 18, 12–16.

Department of Sport, Recreation and Tourism (1986) *Economic Impact of the World Cup of Athletics held in Canberra in October 1985*. Canberra: Research and Development Section of the Department of Sport, Recreation and Tourism.

Downey, B. (1991) The tourism impact on Victoria of its special sporting events: Including case studies of the 1989 Bells Beach Easter Surf Carnival and the 1989 Ford Australian Open. Masters Thesis, Victoria University of Technology.

Faulkner, H. and Raybould, M. (1995) Monitoring visitor expenditure associated with attendance at sporting events: An experimental assessment of the diary and recall methods. *Festival Management and Event Tourism* 3, 73–81.

Gelan, A. (2003) Local economic impacts: The British Open. *Annals of Tourism Research* 30, 406–425.

Gibson, C., Waitt, G., Walmsley, J. and Connell, J. (2010) Cultural festivals and economic development in regional Australia. *Journal of Planning Education and Research*, 29, 280–293.

Gratton, C., Dobson, N. and Shibli, S. (2000) The economic importance of major sports events: A case study of six events. *Managing Leisure* 5, 17–28.

Hall, C. (1992) *Hallmark Tourist Events: Impacts, Management and Planning*. London: Belhaven Press.

Hone, P. (2005) Assessing the contribution of sport to the economy. Working Paper 2005, no. 2. Melbourne: Deakin University.

Howard, D.R. and Crompton, J.L. (1995) *Financing Sport* (2nd edn). Morgantown: Fitness Information Technology, Inc.

Irwin, R., Wang, P. and Sutton, W. (1996) Comparative analysis of diaries and projected spending to assess patron expenditure behaviour at short-term sporting events. *Festival Management & Event Tourism* 4, 29–37.

Lennox Wave, The (2004) *The Lennox Wave*, 8 (July).

Murphy, P. and Murphy, A. (2001) Regional tourism and its economic development links for small communities. In M.F. Rogers and Y.M.J. Collins (eds) *The Future of Australia's Country Towns* (pp. 162–171). Bendigo: La Trobe University, Centre for Sustainable Regional Communities.

Rusty Gromfest (2005) *Gromfest*. On WWW at www.gromfest.com.au.
Taylor, J. (2003) *Mountain Bike World Cup 2002 – Fort William: Economic Impact Study*. Edinburgh: SportScotland.
Tindall, P.A. (2005) Economic impacts of a small-scale youth sports event: A study of the Gromfest Surf Carnival. Honours Thesis, University of New South Wales.
Wall, G. and Mitchell, C. (1989) Cultural festivals as economic stimuli and catalysts of functional change. In G. Syme, B.J. Shaw, D.M. Fenton and W.S. Meuller (eds) *The Planning and Evaluation of Hallmark Events* (pp. 195–202). London: Avebury Press.
Walo, M., Bull, A. and Breen, H. (1996) Achieving economic benefits at local events: A case study of a local sports event. *Festival Management and Event Tourism* 4, 95–106.

Chapter 6
Greening Rural Festivals: Ecology, Sustainability and Human-Nature Relations

C. GIBSON and C. WONG

Introduction

From Ibiza to Byron Bay, festivals have been linked to environmental degradation, a simple consequence of the numbers attending and their immediate impacts on delicate ecosystems. Environmental impacts are linked to the scale of visitation and the capacity of communities to support festivals with appropriate infrastructure. At the same time, festivals can provide unique opportunities for people to come together in the celebration and promotion of environmental causes – whether through festivals with an overt 'green' educational message (Curtis, 2003) or through management practices aimed at reducing per capita consumption of resources. By advocating practices such as recycling, use of public transport, waste minimisation and use of sustainable materials and services, festivals seek to 'green' their image and make practical improvements on their environmental record. Successful 'green' festivals send a powerful place-marketing message to visitors (especially urban ones) that their town or village is forward-minded and switched on to contemporary issues. Festivals also immerse people – literally – within nature, especially rural festivals held on farms, in parks or near scenic landscapes; and festivals bring nature to the people, in the case of agricultural shows, food and wine festivals and flower shows. This chapter discusses how festivals are bound up in questions of environmental impact and sustainability, how rural festivals in particular catalyse new encounters between humans and nature, and what this means for understanding festivals as a 'green' issue.

Environmental Impacts of Festivals

Only quite recently have questions of the environmental dimensions of festivals and event tourism become the focus of discussion, with emphasis previously favouring economic impacts and issues of cultural change and inclusion (Hall, 1989). Even though festival and event managers are now increasingly attempting to make festivals more 'green', it remains the case that peer-reviewed scientific studies of the environmental impacts of festivals are rare. In one study, a festival involved prolonging the use of lights, thus disturbing and delaying the emergence of bats, impacting their feeding habits as well as their ability to raise their offspring (Shirley *et al.*, 2001). Another assessed the noise pollution produced by a festival (Gupta & Chakraborty, 2003). In some instances where particular environmental impacts have been discussed, festivals were not even especially the focus, but rather acted as a background where fireworks or bonfires were discussed for how they release contaminants in the air such as dioxins, heavy metals or suspended particulates (e.g. Dyke *et al.*, 1997; Fang *et al.*, 2002; Farrar *et al.*, 2004; Kulshrestha *et al.*, 2004; Lee *et al.*, 1999; Ravindra *et al.*, 2003). Other studies surveyed residents and visitors about the perceived environmental impacts of a festival, without actual measurement of environmental degradation (e.g. Gursoy *et al.*, 2004), or discussed the environmental implications of permanent facilities built for events such as the Olympics (e.g. May, 1995). This latter issue is relevant for mega or hallmark events but less so for rural festivals as they tend to be small (Gibson *et al.*, 2010), and make use of the existing infrastructure or erect temporary facilities.

That festivals have not been subject to much environmental impact research is not to say that they are trivial or have no negative environmental consequences. Development applications to authorities for permission to stage a festival are becoming more common – especially in Australia and in other countries where environmental regulators such as the Environmental Protection Authority have developed national standards on water and air quality. Because of this, it seems likely that more research will be needed on the environmental impacts of festivals – at the very least so that issues of pollution and waste can be compared across festivals, enabling improvements to be made.

How are Environmental Impacts Measured?

There is no single methodology that will adequately describe and quantify the range of environmental issues generated by festivals. Biophysical techniques, such as soil sampling, water quality testing and ecosystem surveying, could be selected as methodologies to assess the environmental impacts of festivals, but such techniques are unable to quantify impacts that occur beyond the festival site in question. Ecological impacts may extend beyond the physical boundaries of a festival location (through, for example, attendees' transport emissions). Also, biophysical methods are unable to account for the indirect effects of consumption of natural resources (e.g. energy) used to stage the event, or how these vary at dissimilar festivals. Third, although it might be ideal to simultaneously employ several biophysical survey methods, this would be difficult and complicated, requiring much scientific labour as well as being expensive and time consuming. Given that most rural festivals are small, it is no surprise that comprehensive studies of the environmental impacts of festivals are rare.

In the absence of an overall technique, some festivals model the water and energy usage required to operate a festival, usually as part of efforts to contribute to an overall sustainability agenda, and to be able to market a festival as 'green'. Online carbon calculators provide one means at little or no cost (though their accuracy or applicability to all kinds of festivals is not certain). Carbon calculators produce estimates of the quantities of carbon used in staging a festival, and then are able to neatly recommend the purchase of carbon offset credits. Large festivals are also increasingly contracting specialist consultancy firms for more detailed carbon and water use audits (see Chapter 12). Again – there is no consistency across festivals: specialist consultancies tend to bring their own individual expertise, preferred models and measurements to bear on the festival in question and carbon calculators each have their own algorithms.

In the absence of biophysical surveys, a comprehensive whole-of-event measurement of an 'ecological footprint' is a possible alternative (Wong, 2005). Devised by Wackernagel and Rees (1996), the 'ecological footprint' measures the 'load' imposed by a given population on the environment. The ecological footprint documents how much of the annual regenerative capacity of the biospheres (expressed in mutually exclusive hectares of biologically productive land) is required to renew the resource throughput of a defined population in a given year, with the prevailing technology and resource management of that year (Monfreda *et al.*, 2004).

The appeal of the ecological footprint for festivals lies in the fact that various demands on the environment could be examined (e.g. food consumption, resource use, waste disposal and carbon dioxide emissions) in a manner more comprehensive than quick calculator or auditing exercises limited to energy use, while remaining fairly simplistic in its execution.

Because ecological footprints measure the *land area* necessary to sustain current levels of resource consumption and waste assimilation, the results of ecological footprints are also more easily communicated to non-academic audiences (than, for instance, tonnes of carbon dioxide, which people struggle to conceptualise).

Wong (2005) tailored the ecological footprint method for specific application to a music festival, Splendour in the Grass, in Byron Bay, a small but immensely popular town on Australia's east coast with some 5000 people outside festival times. For Splendour in the Grass, three estimates of the overall ecological footprint of the festival (low, medium and high) were produced using data obtained from surveys of festival participants, from actual water and energy-use readings on-site, and from existing consumption models. The festival's ecological footprint was calculated to be 1.12 hectares per capita (or 1.53 global hectares per capita) for the low estimate, 1.43 ha/cap (or 1.75 gha/cap) for the medium estimate and 1.75 ha/cap (or 1.96 gha/cap) for the high estimate. When compared with those of previous applications of the ecological footprint, the findings indicated that an average festival attendee demanded much less ecological space in an aggregate sense than an average resident nationally. In other words, it appears possible that attending a festival is less demanding on environmental resources than staying at home, undertaking 'normal' daily activities around the house. This is because the many forms of energy (e.g. sound and lighting) and inputs consumed by people at a festival, as well as the festival site itself, *are consumed collectively* – by a large number of densely located people and thus consumed at a lower per capita level – than if the same number of people were at home or going about their ordinary business.

Where festival versus national ecological footprint results differed – in the case of the Splendour in the Grass festival – was on the per capita energy required to support travel behaviour: transport resource was much higher, proportionally, for the festival than the national average. Splendour in the Grass is a music festival known nationally as one of Australia's premiere 'alternative' music festivals, and thus a very large proportion of attendees travel great distances to be there (53% travelled distances greater than 1000 km). Making matters worse, the majority of festival attendees (69%) chose to drive to the festival.

Cross-tabulations of variables were also conducted to ascertain relationships that could aid in the prediction of ecological footprints for future festivals. It was found that certain demographic variables (e.g. gender, occupation, industry) were associated with consumption patterns (e.g. choice in food, accommodation, transport). Women tended to eat less food at festivals than men; students were more likely to use 'lower impact' accommodation types such as camping than those in better-paid professional jobs, and used less water accordingly. However, students travelled further: 53% travelled between 1000 and 2000 km to get to the festival site, most from Sydney or Melbourne. Those more likely to come by car to the festival were also more likely to buy souvenirs and other items available at festival stalls (thus increasing per capita overall consumption levels). Festival audiences are not a homogenous mass, but rather a diverse community who consume – and thus impact on the environment – in complex ways.

Beyond 'Impact': Festivals and Human-Nature Relations

Measuring human impacts on environment – through tools such as biophysical testing and ecological footprint analysis – makes it possible to tell whether festivals damage local ecosystems, and to estimate how a range of resources are consumed in staging festivals. From this it is possible to make useful interventions in festival management in order to reduce environmental harm.

However, as geographers such as Sarah Whatmore (1999), Noel Castree (2002) and Lesley Head (2008) have shown, there are particular practical and conceptual problems with adopting an approach focused on human *impacts on* the environment. Human-impact research relies on an exceptionalist assumption that humans are distinct from nature (and impacting negatively on it) rather than seeing humans as different from other plants and animals, but still very much a part of nature – one of many agents in interactive ecological systems, forging connections, growing dependencies and indeed, causing much damage. This critique of the human–nature binary underpinning 'impact' is important, because conceptual understanding of humans as separate from, but acting upon, an inert non-human nature, have resulted in unhelpful practices that can ironically *prevent* improvements in environmental management from taking place. Breaking down the assumption of a passive nature impacted upon by destructive humans enables consideration of longer-held negotiations of environmental conditions by Indigenous peoples. It also reveals more complex and iterative processes of interaction and adaptation between humans and non-human nature – from which important ethical, conservation and biodiversity lessons can be learned.

Following this line of critique, new ways of framing festivals as cut through by 'environmental' issues are possible. Beyond measuring the direct impacts of festivals on the environment, it becomes possible to discuss how festivals catalyse particular encounters between humans and nature – how festivals reconfigure human–nature relations, even if temporarily (Gibson, 2010). From considering these kinds of encounters, it is possible to rethink what the 'greening' of festivals might mean more broadly.

This is especially the case in rural festivals, where encounters with 'nature' may be the rationale for an event, a key reason for hosting the festival in the first place, or a unique selling point compared with metropolitan events. The oldest of all festivals – pagan rituals, harvest celebrations, equinoxes, Aboriginal ceremonies – were much like this, commemorating the evolving rhythms of nature, marking the turn of the seasons. Traces survive in such festivals as agricultural shows (timed to take advantage of regional harvests; see Chapter 2 by Darian-Smith), cherry blossom festivals (in Japan) and full-moon parties. In Australia, modern festivals revive such practices: both Tumut's Festival of the Falling Leaf and Bathurst's Autumn Heritage Festival celebrate the turning colours of deciduous trees (in a country mostly covered by evergreens); Katoomba's Winter Magic Festival and Hobart's Antarctic Midwinter Festival raise spirits in the depths of their cold-climate winters; while the Oberon Daffodil Festival of Spring Gardens (in Blue Mountains, NSW) and the Australian Springtime Flora Festival (the largest gardening festival in Australia, in Kariong, NSW) are among spring bloom flower festivals too numerous to list here. All rely on the graceful turn of seasonal nature, and create communities around benevolent encounters with the weather, the seasons, plants and animals.

At some festivals, nature might be a *raison d'être*, but human regard for nature may not necessarily be reconfigured substantially, any further than, for instance, a general appreciation of nature's beauty and its aesthetic appeal. A good number of music festivals held on rural properties might fit into this category: appealing to an Arcadian idyll, to a generic, escapist sense of pastoral retreat, that is, nature as a nice backdrop to a party. Debates about the ecological impact of trance and techno festivals (when staged in forests or on scenic pastoral properties) have involved such a critique – pointing out the contradiction that techno festivals are often more about psychotropic adventures of the mind, than a commune with nature – and can actually produce levels of environmental damage through trampling, light and noise pollution, that undercut any gains in ecological awareness. At festivals non-human nature might be commodified outright, with less concern for transforming the environmental consciousness of participants than for emptying their wallets.

But in other cases, festivals centred around nature are indeed connected to quite overtly transformative agendas, such as the Yellow Gum Winter

Flowering Festival in Bannockburn, Victoria, a festival aimed at celebrating, in the words of its flyer, 'the annual winter flowering of the local Yellow gum *Eucalyptus leucoxylon* ssp *connata*, which is an important source of food and shelter for many resident and migratory birds and mammals'. The Yellow Gum Winter Flowering Festival complements year-round efforts by the Friends of Bannockburn Bush landcare group to revegetate weedy soil dump sites, and aims to raise awareness of ecological inter-connectivity as exemplified by the gum's annual flowering. At the festival, for instance, guided walks and talks by landcare volunteers draw attention to the plant and its role in bird migrations. Indeed, not merely is the festival aimed at raising environmental awareness among humans, its organisers even claim on its festival website that 'the nationally significant Swift Parrot comes all the way from Tasmania to be at our Festival' – humans and nature entwined.

In Parkes, NSW, the site of an internationally significant research telescope (known colloquially as 'The Dish'), the rhythms of celestial nature provide a different source of inspiration, for the annual AstroFest, organised by the Central West Astronomical Society. According to organisers, 'The Central West of NSW boasts some of the darkest skies in Australia, and as such is a wonderful place to appreciate the wonders of the night sky … . The festival endeavours to bring to the people of the Central West world renowned astronomers, both professional and amateur, so they may share their enthusiasm and love of the heavens' (CWAS, 2010: 1).

At both Inverell's Opera in the Paddock and Bermagui's Four Winds Festival, the opportunity to see and hear classical music performed in the 'natural' setting is a major reason for attendance, above and beyond the actual music. 'Natural' landscapes can provide the ideal 'frame' for classical music, as in Bermagui's tranquil waterside festival site, or provide striking juxtaposition, as in South Australia's Opera in the Outback. In an example of ultimate juxtaposition, SnowFest in Gloucester, New South Wales involves the complete importation of snow from further afield, that is, imported nature in an otherwise decidedly unsnowy environment (see Chapter 4).

Festivals are therefore capable of providing opportunities to engage people with nature in all manners; some playful, others more deeply committed to conservation. Festivals can promote environmental messages in ways that are fun, creative or experimental. Some focus on environmental restoration activities, such as the Mount Elephant Festival in rural Victoria, which involves music, stalls and actual restoration efforts, on an ex-farm site that local residents rallied around converting to publicly owned biodiverse habitat. According to organisers, local community groups raised money to buy the farm in 2000 from its previous owners, and staged an annual festival to raise further funds to revegetate the site and to raise environmental awareness about local habitat (Anon., 2003). Even when in some years rain affected attendances (and subsequent fundraising activities), the festival was still considered 'worth it' for its contribution to the site's conservation and educational values.

Indeed, festivals can reach audiences otherwise rarely exposed to conservation or sustainability ideas. Not everybody ranks environmental issues as most concerning to them (even people who might be 'pro-environment' at some level) and certainly, only a minority of the general public are actively involved in environmental education, restoration or activism. With roots in community affairs and frequently drawing in the support of schools, community choirs, amateur theatre groups and Rotary and Lions Clubs, festivals can provide ways for communities to celebrate their local and wider environments, without having to resort to green dogma or rely too heavily on 'expert' knowledges (Curtis, 2003). But also, the ever-present field of stalls at festivals provides space for environmental scientists and non-profit organisations to communicate to a general public audience, especially so at festivals with a nature/environment theme. At the Port Stephens Whale Festival in NSW stalls from the Organisation for the Rescue and Research of Cetaceans (ORRCA), Ocean and Coastal Care Initiative, Conservation Volunteers Australia, Hunter Wetlands Centre, and Marine Mammal Research are all present. As well as live entertainment, the festival also features daily mock up whale rescues (Port Stephens Whale Festival, 2009) – a particularly vivid example of how to engineer festival encounters with nature.

Elsewhere on the same coast, Eden's Whale Festival features whale-watching tours, a parade in the theme of 'whales, mermaids and neptunes' and tours of the town's historic whaling station. What makes this festival significant is that it takes place in a previously iconic whaling town. The festival thus cements Eden's economic transformation and updated place identity: once a town with a predatory view towards nature, Eden (as exemplified with its festival) fosters a new relationship with whales based on preservation and coexistence.

Such festivals can be large in scope and ambition: in the case of the Tweed River Festival in NSW, official events spanned a full month – dedicated to celebrating an entire river catchment. The festival included a Biodiversity in Art exhibition; foreshore/wetland and bird information walks; an open day at a special purpose Sustainable Living Centre ('learn how to reduce your carbon footprint and live more sustainably'); a launch of a DVD about attempts to restore and rehabilitate the banks of the Tweed

River; round-table meetings of catchment managers 'to share knowledge about natural resource management in the Tweed'; guided kayak tours; the Tweed River Classic Boat Regatta ('lovingly restored or re-created classic boats cruise the Tweed River'); dragon boat races; river swim races; live entertainment and stalls dedicated to catchment environmental issues. Such events promote environmental awareness – of a catchment, of a local plant, of a celebrated local natural feature – amidst activities that are not otherwise always about the environment, per se. Through music, stalls, rides, games, competitions, food and incidental interactions of people of diverse ages and backgrounds, an air of conviviality descends on festivals that in turn can create a suitably open-minded context within which to engage with environmental themes.

There is then, something about the carnival atmosphere at festivals, which enables human–nature relations to be reconfigured in creative and engaging ways. At the Garma Aboriginal festival, held annually since 1999 by the Yothu Yindi Foundation on the Gove Peninsula in remote north east Arnhem Land, the festival format creates an arena in which tribal Yolngu celebration of the country and connection to land is reproduced, *and* in which other non-Aboriginal visitors are invited to share in that sense of celebration and reverence for human–nature bonds (see Chapter 7). At another Aboriginal festival, the Yaamma Festival in Bourke (western NSW), visiting Indigenous artists and local school children install artworks along the banks of the Darling River, as part of aims to 'unite communities along the river system and promote healing and reconciliation between all peoples' (Outback Arts Inc., 2010). Here, as at Garma, festivals enable Indigenous knowledge about nature to be practised and renewed, and to circulate throughout the wider community.

'Productive' Nature

In contrast, other (often more traditional) festivals have twisted human–nature relations in less emancipatory ways, celebrating mastery over nature, the success of the colonial project and ability to productively convert settled country into industrial production. This has long been the case at agricultural shows (Anderson, 2003) which throughout Australian history have advanced modernist ideals of technological improvement, masculine rurality and the submission of nature to human ingenuity (see Chapter 2). But, plagued by competition from other entertainment forms and the struggle to remain relevant in multicultural, post-industrial Australia, many agricultural shows have in recent years merged to form larger, generic regional events, restructuring their programs so that events celebrating mastery over nature (such as woodchopping and livestock judging) are only one among a range of newly introduced attractions including live music, fairgrounds, dog shows, stunt cars and motorbikes and showbags of candy and toys for children. Some regional shows now reflect wider transformations in the Australian agricultural industries, shifting their focus away from the celebration of rural bounty towards niche agricultural production, including organic food, gastronomy, celebrity cooking and gardening. Modernist celebration of the mastery of Australian industrial farming is increasingly supplanted by sophisticated marketing of post-Fordist, foodie culture, itself suggesting new ways of commodifying and selling nature.

Beyond agricultural shows, other festivals of food have sought to celebrate regional agricultural, viticultural and horticultural products and in doing so are generating new 'cultures of nature' (Castree, 2005: xxii) across rural Australia. Diversification of Australian agricultural production is reflected in celebrations of increasingly diverse food types. Competitive natural advantage in agriculture becomes a means to construct new place identities, as niche market production becomes further specialised and new natures come to be associated with regions. Here are but a few examples (from Gibson *et al.*'s (2010) database of over 2800 rural festivals in Australia): the Batlow Apple Blossom Festival, Casino Beef Week, Collector's Pumpkin Festival, the Guyra Lamb and Potato Festival, Tweed Valley Banana Festival, Pakenham's Celebrate Asparagus festival, and an annual Nut Festival in the tiny alpine settlement of Wandiligong: 'In April and May, walnuts and chestnuts fall from the trees in orchards, backyards and along the lanes of Wandiligong where they are gathered by pickers, residents and visitors. The Wandiligong Nut Festival celebrates this harvest by hosting a two-day festival' (Wandiligong Nut Festival, 2010: 1).

Metropolitan tastes, cultural capital and nature again combine in the proliferation of produce festivals throughout rural Australia (where regional wineries, landscapes, primary produce and even locally distinct soil types feature in the rationale for festivals), as well as festivals dedicated to organic food, slow food and seafood – often playing on region-specific production and niche marketing of wine (Tenterfield, NSW), olives (McLaren Vale, South Australia), cheese (Hunter Valle, NSW), beer (Maitland, NSW) and oysters (Ceduna, South Australia). Multiculturalism too refracts engagements with nature. In Woolgoolga, NSW (a small coastal town unusual for its large Indian migrant community), a curry and chilli festival has become a highly popular drawcard for the entire north-coast region. The migration of Igor Van Gerwen, a Belgian master confectioner, to Latrobe, Tasmania led to investment in a local chocolate factory, and enabled that town to put on a chocolate festival – held in the middle of the winter in the coldest Australian state (itself relying on frigid nature in a way barely possible anywhere else in the country for fear of the chocolate melting). In Broome, the Shinju Matsuri Festival 'rekindles the excitement and romance of Broome's early days as a world-renowned producer of Pearls and Pearl shell (Shinju Matsuri, 2010), but has also become that remote town's key arts festival and celebration of Indigenous culture and migrant history (the Japanese, Chinese, Malay, Koepangers and Filipinos were all involved in pearl harvesting and diving, as were Indigenous people).

There is a corollary too in flower shows and gardening festivals – that once celebrated the domestication of nature and reproduction of English gentility in the colonies (Chapter 9), but are now increasingly reflective of multicultural 'ways of being with nature', incorporating native flowers (as at Hobart's Floral Festival, which has as its overt aim the promotion of native planting) and gardening styles from the Mediterranean, Bali and Singapore. These are complemented by specialist festivals for all manners of gardening enthusiasts and collectors: festivals are now dedicated to daffodils (Braidwood and Oberon, NSW), roses (Goulburn, NSW, Cavendish, Morwell and Skipton, Victoria), camellias (Hobart), tulips (Wynyard, Tasmania, Silvan, Victoria), irises (Hamilton, Tasmania, Rainbow, Victoria), dahlias (Portland, Vic), orchids (Warrnambool, Wangaratta and Ballarat, Vic), rhododendrons (Blackheath), chrysanthemums (Bendigo), geraniums (Horsham, Vic) and begonias (Ballarat). In one of the more over-the-top examples, no less than half a million bulbs are planted for the Tesselaar Tulip Festival in Silvan, Victoria:

> Tulips are magic for so many reasons, their exotic majesty a reflection of nature's miraculous beauty. They encompass every colour from dazzling to delicate hues in patterns most daring and others most plain. They have inspired poets and artists and bewitched nations with their allure. They are the stuff of legend, and rumoured to be the sleeping cradles of the fairies. They are even said to have their origin in the blood of a star crossed lover … . Escape the every day for a little hocus pocus at a festival packed with entertainment and splendor. Delight in the magic of spring at the Tesselaar Tulip Festival. (Tesselaar Tulip Festival, 2010: 1)

Whether framed as gastronomic tourism, industrial bounty or specialist obsession, productive nature features and is recast in myriad ways in rural festivals.

Conclusion

Festivals cannot help but impact on the environment. They bring people, cars, noise and generate waste. Tracks and fields are trampled, the air polluted (especially through carbon dioxide from cars) and local transport, sewerage and emergency services are stretched. These can be particularly acute in the case of rural festivals, given the small scale of host towns (and their infrastructure) and sensitivities regarding proximate ecosystems, habitat and national parks. Methods have been proposed that seek to model these impacts – such as the ecological footprint method – and private sector waste management firms are developing increasingly sophisticated methods of collecting recyclable materials (thus minimising landfill). But much more could still be done, especially to reduce car dependency (given it appears to be by far the largest component of the overall environmental impact at festivals). Also still lacking is a shared stock of knowledge and experiences of festival environmental management. At present, understanding the environmental impacts of festivals still tends to be approached on a case-by-case – and largely ad hoc – basis. It remains difficult to tease apart substantively 'successful' examples of environmental management at festivals from marketing greenwash, and there are too few scientific studies that adequately model total environmental impacts, meaning there is no clear-cut, accepted method of analysis.

Beyond immediate impacts on air, water, energy use and waste, festivals (and rural festivals especially) need to be understood as moments where it is possible for people to encounter nature collectively, and intensely. In some cases, such encounters prove controversial, to the point that festivals have to find alternative accommodation (such as the Splendour in the Grass festival in Byron Bay, whose application to locate on rural property adjacent to koala habitat was controversial, leading to it moving to a different state; see Gibson & Connell, forthcoming). But encounters with nature need not be so polemical: other (often smaller, non-profit) events celebrate landscapes without mass destruction, and even incorporate environmental restoration activities. Some festivals construct nature as a 'wild' force successfully tamed by humans (in the case of agricultural festivals), as ephemeral beauty thence controlled and made domestic (flower shows) or appreciated in new, and more benign ways (whale festivals). Festivals generate new environmental knowledge (e.g. raising awareness of the totality of a river catchment, in a region otherwise fragmented into scattered villages and towns) and echo old ones (in the case of Aboriginal love of the country). 'Greening' festivals requires far more than mere introduction of recycling initiatives or purchasing carbon offsets: festivals – and especially rural festivals – are opportunities to rethink the way in which we relate to non-human nature, and reconfigure our practices accordingly.

References

Anderson, K. (2003) White natures: Sydney's Royal Agricultural Show in post-humanist perspective. *Transactions: Institute of British Geographers* 28, 422–441.

Anon. (Mt Elephant festival organiser), interviewed by Chris Gibson. 1.12.2003.

Castree, N. (2002) Environmental issues: From policy to political economy. *Progress in Human Geography* 26, 257–365.

Castree, N. (2005) *Nature*. London: Routledge.

Central West Astronomical Society (CWAS) (2010) CWAS AstroFest. On WWW at http://www.parkes.atnf.csiro.au/news_events/astrofest/ Accessed 27.06.10.

Curtis, D.J. (2003) The arts and restoration: A fertile partnership? *Ecological Management and Restoration* 4, 163–169.

Dyke, P., Coleman, P. and James, R. (1997) Dioxins in ambient air, bonfire night 1994. *Chemosphere* 34, 1191–1201.

Fang, G-C., Chang, C-N., Wu, Y-S., Yang, C-J., Chang, S-C. and Yang, I-L. (2002) Suspended particulate variations and mass size distributions of incense burning at tzu yun yen temple in Taiwan, Taichung. *The Science of the Total Environment* 299, 79–87.

Farrar, N.J., Smith, K.E.C., Lee, R.G.M., Thomas, G.O., Sweetman, A.J. and Jones, K.C. (2004) Atmospheric emissions of polybrominated diphenyl ethers and other persistent organic pollutants during a major anthropogenic combustion event. *Environmental Science and Technology* 38, 1681–1685.

Gibson, C. (2010) Geographies of tourism: (Un)ethical encounters. *Progress in Human Geography* 34 (4), 521–527.

Gibson, C. and Connell, J. (forthcoming) *Music Festivals and Regional Development*. Ashgate: Aldershot.

Gibson, C., Waitt, G., Walmsley, J. and Connell, J. (2010) Cultural festivals and economic development in regional Australia. *Journal of Planning Education and Research* 29, 280–293.

Gupta, A. and Chakraborty, R. (2003) An integrated assessment of noise pollution in Silchar, Assam, North East India. *Pollution Research* 22, 495–499.

Gursoy, D., Kim, K. and Uysal, M. (2004) Perceived impacts of festivals and special events by organizers: An extension and validation. *Tourism Management* 25, 171–181.

Hall, C.M. (1989) Hallmark tourist events: analysis, definition, methodology and review. In G.J. Syme, B.J. Shaw, D.M. Fenton and W.S. Mueller (eds) *The Planning and Evaluation of Hallmark Events*. Aldershot: Ashgate.

Head, L. (2008) Is the concept of human impacts past its use by date? *The Holocene* 18, 373–377.

Kulshrestha, U.C., Nageswara Rao, T., Azhaguvel, S. and Kulshrestha, M.J. (2004) Emissions and accumulation of metals in the atmosphere due to crackers and sparklers during Diwali Festival in India. *Atmospheric Environment* 38, 4421–4425.

Lee, R.G.M., Green, N.J.L., Lohmann, R. and Jones, K.C. (1999) Seasonal, anthropogenic, air mass, and meteorological influences on the atmospheric concentrations of polychlorinated dibenzo-p-dioxins and dibenzofurans (Pcdd/Fs): Evidence for the importance of diffuse combustion sources. *Environmental Science and Technology* 33, 2864–2871.

May, V. (1995) Environmental implications of the 1992 Winter Olympic Games. *Tourism Management* 16, 269–275.

Monfreda, C., Wackernagel, M. and Deumling, D. (2004) Establishing national natural capital accounts based on detailed ecological footprint and biological capacity assessments. *Land Use Policy* 21, 231–246.

Outback Arts Inc. (2010) Building creative communities. On WWW at http://www.outbackarts.com.au. Accessed 27.06.10.

Port Stephens Whale Festival (2009) 2009 Port Stephens Whale Festival. On WWW at http://www.whalefest.com.au. Accessed 27.06.10.

Ravindra, K., Mor, S. and Kaushik, C.P. (2003) Short-term variation in air quality associated with firework events: A case study. *Journal of Environmental Monitoring* 5, 260–264.

Shinju Matsuri (2010) Celebrating 40 years of Shinju Matsuri. On WWW at http://www.shinjumatsuri.com.au/history.html. Accessed 27.06.10.

Shirley, M.D.F., Armitage, V.L., Barden, T.L., Gough, M., Lurz, P.W.W., Oatway, D.E., South, A.B. and Rushton, S.P. (2001) Assessing the impact of a music festival on the emergence behavior of a breeding colony of Daubenton's bats (*Myotis Daubentonii*). *Journal of Zoology* 254, 367–373.

Tesselaar Tulip Festival (2010) Tesselaar Tulip Festival: Discover tulips. On WWW at http://www.tulipfestival.com.au. Accessed 27.06.10.

Wackernagel, M. and Rees, W. (1996) *Our Ecological Footprint: Reducing Human Impact on the Earth*. Philadelphia: New Society Publishers.

Wandiligong Nut Festival (2010) Here we go gathering nuts in May: May 8th and May 9th 2010. On WWW at http://www.brightvic.com/wandinutfestival. Accessed 27.06.10.

Whatmore, S. (1999) Culture-nature. In P. Cloke, P. Crang and M. Goodwin (eds) *Introducing Human Geographies* (pp. 4–11). London: Arnold.

Wong, C. (2005) The environmental impacts of a festival: Exploring the application of the ecological footprint as a measuring tool. Honours thesis, University of New South Wales.

Part 3
Politics and Place: Culture, Nature and Colonialism

Chapter 7
Performing Culture as Political Strategy: The Garma Festival, Northeast Arnhem Land

P. PHIPPS

This chapter discusses Australia's most prominent Aboriginal festival, a festival which takes place in Arnhem Land in Australia's far north, and to which great cultural and political significance is attached. Aboriginal peoples and cultures are a prominent 'artefact' in non-Indigenous Australian public discourse and culture. Characteristically these discourses are concerned with the very real disadvantages in health, education, employment and life expectancy faced by Aboriginal people, sometimes expressed as moral panic; while occasionally they romanticise or mystify Aboriginal peoples and cultures. Neither position – abjection or romanticism – really engages with the experiences or aspirations of Aboriginal people themselves. Most of the Australian population (who live in large cities) have had minimal or no contact with Aboriginal people, or fail to recognise or engage with the Aboriginal people living and working around them. As a small and very diverse minority of the national population, Aboriginal people have deployed a whole range of strategies to further their twin objectives of equality of opportunity on the one hand (after a long history of severe discrimination); and on the other hand the recognition of special rights to pursue and maintain their distinctive cultures and connections to their ancestral lands despite a long history of pressure to assimilate and disappear.

The experience of colonialism has differed enormously from the more densely and early-colonised parts of the country, where Indigenous languages, religions and peoples' relationship to land were severely repressed; to those more remote places, particularly in the centre, north and west, where Aboriginal labour was needed, and difference less-thoroughly oppressed. It was not until the middle of the 20th century that Australian colonial institutions really took hold of daily life in Arnhem Land (a remote, region in the northeast of what is today the Northern Territory, set aside in the 1930s as a vast Aboriginal reserve). Yolngu people – the traditional Aboriginal people of Arnhem Land – were then encouraged to live under the authority of missionaries in centralised towns such as Yirrkala, Galiwin'ku, Ramingining, Maningrida and others. The Aboriginal rights-based initiatives of the 1960–1970s (equal pay campaigns, citizenship, land rights struggles), and the social dislocation elders saw in the towns, led many Yolngu to establish themselves back on *country* (ancestral lands) in what came to be known as the homelands movement. As a result of their relatively late and 'incomplete' colonial experience, Yolngu people have been prominent in these movements and still maintain strong connections to their ancestral lands and the law, language, spirituality, dance and music which sustain it. In this chapter, I discuss one event where these connections to country are vital – the Yolngu Garma Festival.

Christie and Greatorex (2006) describe a pattern of habitation where today's population of about 5000 Yolngu are distributed and moved between these former mission towns of 500–2000 people, small homeland settlements, and about 500 people in the regional capital Darwin for medical, educational, social or other reasons. The maintenance of these connections to country is the main concern of Yolngu ceremonial life, most active for frequent and demanding funerals. These connections have also generated a dynamic Yolngu visual arts sector generating income through community arts centres' linkages to a national and global art market. Similarly, ranger programs translate Yolngu land-management practices into an idiom recognisable (and fundable) by the Australian state, and Yolngu educators have been struggling to do the same for over two decades. Yolngu have forged a small world of Yolngu modernity where their community organisations and corporations mediate relations with state bureaucracies and the mine; while ceremonial life, hunting, gathering and associated cultural transmission continue and are transformed. The Garma festival is one such site of transmission and transformation.

The Garma Festival

Garma is an intercultural gathering of national significance, and simultaneously is a local gathering of Yolngu clans on Yolngu land for Yolngu political, ceremonial and recreational purposes. Yolngu landowners invite visitors to participate in a five-day cultural event, held annually since 1999 by the Yothu Yindi Foundation. The Garma site is relatively accessible to the well-connected Gove airport as a result of the substantial open-cut bauxite-mining infrastructure, which has dominated the region for the last 40 years. Garma typically involves about two thousand participants gathering at this site of temporary bush-pole and portable administrative and catering buildings, bush shelters, makeshift shower blocks, toilets and over a thousand domed tents making up Yolngu family-based camps and the concentrated mass of visitors' 'tent city'; all organised around a central, sand-covered ceremonial ground.

Over a decade, the Garma Festival has built up a suite of activities and programs designed to engage a range of different visitor constituencies physically, intellectually and spiritually, while bringing benefits to the Yolngu host community. These multiple constituencies are a complex mix including: Yolngu hosts the Gumatj clan and their intimate Rirratjingu 'mother' clan, other Yolngu clans and organisations, other Indigenous people from Australia and elsewhere, *Balanda* (non-Indigenous visitors), Yolngu and Balanda youth from the region, tertiary students and academics, government and related policy-makers, cultural tourists (divided into men's and women's activities), international *yidaki* (didjeridu) students, media crews, 'VIPs' (including philanthropic, corporate and government sponsors), high school music students from Indigenous rock bands across the NT, '(mining) town visitors', celebrities, staff, and volunteers. The eclectic mix of participants is one of the features that has given Garma a reputation for both intimacy and influence: Federal and Territory Government Ministers rub shoulders with Yolngu elders and artists, the occasional media celebrity or rock star queues for food with networking academics, smiling mining executives carry spears they made under the direction of Yolngu rangers while barefoot kids kick a football in the sand.

Garma carries the intertwined pragmatic purposes of local cultural survival, renewal and resource gathering on the one hand, and visionary local and national cultural transformation on the other. At its simplest Garma was originally seen by its Yolngu founders as a way to advance Yolngu education, training and employment guided by traditional law. Garma grew from a 1998 workshop at Gulkula which was to be the site of a Yolngu 'bush university' or the beginning of an integrated cultural studies education facility (the Garma Cultural Studies Institute) with the original ambition being the construction of culturally appropriate office and service buildings, connected with other organisations like ranger programs, men's and women's healing, a cultural resource centre, and so on. Clan leaders selected the site of Gulkula both for its spiritual connection with the important Yirritja ancestral figure Ganbulabula, and as the former site of Dhupuma College boarding school, a high-school education delivered specifically for Yolngu with great effect in training Yolngu teachers, health workers and other professionals until it was shut down by the Northern Territory Government in 1981 (Gaykamangu et al., 1999; McMillan, 1999).

Much more important than buildings, Garma has made a 'festival' space where formal training in cultural tourism, musical performance and recording, and event security has been happening simultaneously with nationally significant Indigenous policy discussions, while a women's Yolngu traditional healing program has leveraged the Garma 'VIP' network to seek philanthropic support. While Yothu Yindi Foundation Chairman Galarrwuy Yunupingu once asserted that he would prefer to do away with the term 'festival' to make the event simply and uniquely 'Garma', the use of the festival concept is a widely familiar cultural form that provides licence for framing experiences that cross entrenched cultural limits and personal habits. By holding a 'festival' the Yolngu hosts of Garma can muster a range of influential participants who are prepared to forego their usual urban privileges and comforts: queue for food and toilets, sleep in the bush, be out of mobile phone coverage and listen to unfamiliar speaking voices and music through the day and as they sit in the sand watching the evening *bunggul* (ritual dance/corroboree). After dark the school bands that have been in professional mentoring throughout the days leading up to Garma gets their chance to perform their repertoire ranging from rock to hip-hop on the bauxite mound stage. With professional staging and sound production they look and sound their best, and their families and communities come to support them as they play alongside local and regional legends like Yothu Yindi and Saltwater Band, and the national and international acts Yothu Yindi's international fame and networks attract.

Garma is a pragmatic strategy for reinforcing and strengthening local cultural practices, building new resources, and engaging and incorporating influential people in key institutions such as, media, law, health, public administration and education into relationships of reciprocity. This reciprocity opens recognition of forms and practices of Aboriginal sovereignty simply by being there on Yolngu land learning from Yolngu. Galarrwuy Yunupingu is quoted by actor Jack Thompson (Yothu Yindi Foundation, 2002) describing the respectful, cross-cultural learning experience of Garma on Yolngu lands as, 'a vision of Australia as it might be'.

Garma as Political Strategy

The decade of Australian conservative national government from 1996 to 2007 saw a cultural and political stalemate in Australian Indigenous affairs, which elicited many civic responses and initiatives focused broadly on 'reconciliation'. The conservative Howard Government refused to engage with the challenges of reconciliation with Indigenous Australians and its reactionary attack on Indigenous institutions, land rights and vilification of Aboriginal people themselves was hostile to say the least. Despite this, there has been a substantial constituency of the Australian public and key social institutions who remained deeply sympathetic to the notion of Aboriginal self-determination, cultural survival and further progress in formal and informal processes of reconciliation as the most effective ways to remedy Indigenous disadvantage. Among other initiatives the 'Sea of Hands,' and the mass bridge-crossing 'Walks for Reconciliation' in 2000 demonstrated this support on a massive scale, and the possibility of the mutual human feeling that the work of decolonisation requires as a starting point.

This decade offered no signs of hope or comfort for Indigenous interests. It was a case of late-colonial business as usual, with a deeply conservative government setting policy directions inimical with Indigenous cultural maintenance, social or political rights. Indigenous organisations and leaders used whatever strategies they had available to maintain and advance the political gains they had made in the previous two decades. Land Councils ensured a relatively secure base for Indigenous land management and advocacy, and ATSIC still appeared to be a secure national Indigenous institution. The long-standing Yolngu Chairman of the Northern Land Council (NLC), Gumatj clan leader Galarrwuy Yunupingu, had been a prominent advocate for Yolngu and broader Aboriginal rights for almost thirty years. His skilled English and bicultural understanding qualified him as translator for his father and other plaintiffs in the Gove Land Rights Case (*Millipurm v Nabalco* 1971) seen as a watershed moment in the Land Rights movement (Williams, 1986).

In 1999, Galarrwuy Yunupingu established the Garma Festival of Indigenous Culture with his equally famous brother Mandawuy, lead singer of the popular rock band Yothu Yindi. The Yunupingu brothers, both separately recognised as 'Australian of the Year,' mobilised this unique cultural-political initiative under the organisational structure of the Yothu Yindi Foundation (YYF), supported by a shifting alliance of Yolngu clan groups, principally the Gumatj-Rirratjingu. These two brothers were uniquely well-qualified and resourced to bridge the chasm of mainstream Australian ignorance of Indigenous realities and to make a cultural leap in the process of decolonisation. Over the years Mandawuy Yunupingu's vibrant creativity and Galarrwuy's political momentum, both drawing on a very strong grounding in Yolngu cultural life and law, had gathered together a well-connected network of talent and support from across Australia, and a reservoir of goodwill particularly amongst educated urban 'southerners'. Mandawuy, with a university degree from 'down south,' had been the first Indigenous school principal in Australia (at the bilingual Yirrkala Community Education Centre) and his bi-cultural rock band had an international following which broke into the Australian mainstream with the hit-song *Treaty* in 1992.

Key members of the Yothu Yindi band became central to Garma: Mandawuy as Secretary of YYF and host of the academic and policy Key Forum; Witiyana Marika coordinating and mobilising *bunggul*; the bands' manager Alan James became the CEO of the Yothu Yindi Foundation and Director of the Garma Festival; guitarist Stuart Kellaway ran the music training workshops and became a music teacher at the Yirrkala school with support from other band members. From the start YYF was very good at bringing in outside expertise and connections: strong links with Melbourne, Darwin, Sydney and Canberra-based academics; a former Keating adviser working the media strategy and political connections; strong links across national Indigenous leadership and networks; film star Jack Thompson ever-ready to deliver the Garma message to camera or assembled guests, as well as an assortment of other public icons adding their own charismatic style. In addition long-term collaborators include local educators, local ALP politicians, the art centre, ranger program and school, the NLC, and the community relations manager from what was the miner Nabalco (then Alcan, and now Rio Tinto), with many others providing varying degrees of consultation, support and sometimes contradictions and tension. These kinds of human and institutional resources add up to substantial political clout, but in the Howard years, despite visits from the occasional Minister, even that was not enough to break through the impasse. Garma however, did find a way around the federal political impasse by working at a cultural level, both with important individuals and at the broader level of public discourse and representations.

Garma has been a skilful Yolngu strategy to keep Indigenous issues on the national agenda through a highly localised, very specific, public intervention in the realm of representation and knowledge exchange and production. For the most part it is very effective, evidenced by the stream of Territory and national politicians and policy-makers who appear and speak at Garma, announce policies and initiatives, hand out awards, or just make sure they are present. However, this strategy of exercising political capital is not without its risks. The June 2007 Commonwealth 'Emergency Intervention' into Northern Territory Aboriginal communities (Commonwealth of Australia, *Northern Territory Emergency Intervention Act*, 2007) was deeply disturbing for the affected Indigenous communities, their leaders and Indigenous rights activists nationally. In the absence of a national forum like the now-abolished ATSIC, the national Indigenous leadership turned to Garma as a forum to discuss the intervention and formulate a united response. Galarrwuy stood shoulder to shoulder with the mainstream of the Indigenous response, strongly condemning the intervention at Garma. Noel Pearson kept away from Garma with his qualified support for the plan, a position shared by Garma stalwart Professor Marcia Langton. Two weeks after Garma, Pearson and Langton brought the crusading intervention leader, Indigenous Affairs Minister Mal Brough, for private meetings with Galarrwuy on his homeland.

The prominence of Garma and its Gumatj clan hosts in Indigenous affairs made the prospect of a deal with Galarrwuy worth special compromises by the Minister. Encouraged by Pearson and Langton, and offered an exceptionally sweet development deal for his Gumatj homeland of Gunyangara (Ski Beach) without loss of control over land, Galarrwuy broke ranks with the national Indigenous leadership mainstream and signed on to the intervention. This divisive change of heart not only undermined Galarrwuy's national Indigenous authority, but the authority of Garma as a national Indigenous forum. Indigenous leader Professor Mick Dodson was quoted in *The Age* newspaper (Chandler, 2007) as saying,

> The problem I have is that this doesn't appear to be a sound public policy approach – reacting to criticism in this way. It's bad policy ... The precedent is now set. Jump up and down, and the Government will come in and bring some prominent Aboriginal people who agree with them to talk to you and to do a deal with you to keep you quiet. Is that how it works? Galarrwuy has been one of the most strident and outspoken critics of the intervention, particularly this aspect of it – the leases ... It must be a large inducement to turn his view around.

Within a few months the Coalition Government was swept from power leaving the special Gunyangara deal uncertain if not dead, but the NT intervention policy of the previous government has been broadly kept in place. By July 2008 the first Aboriginal community to host a Rudd Labor Government 'community cabinet meeting' was Yirrkala, whose leaders had taken a consistent stand against the intervention. As one of the regional Yolngu clan leaders present, Galarrwuy took a leading role in welcoming them.

In Arnhem Land and other remote northern Australian regions the land rights movement has been about more than the legal recognition of Aboriginal land ownership. At its most developed in the homelands movement this struggle has been for the reassertion of Aboriginal life and law on *country* (clan-based homelands). This is an emphasis on resuming and renewing local practices while managing the social (rather than material) technologies of modernity with caution. These movements are calling for 'mainstream' (dominating culture) understanding of Indigenous difference to allow an Indigenous modernity to develop relatively free from the intensely destructive stresses of colonial domination. Garma has been a significant instrument in this call.

Performing Garma

The core activity of Garma is the *bunggul* held each evening on the central ceremonial ground at the Gulkula site. These ritual performances depend on the negotiation of Yolngu knowledge-holders, and may include clans from far away, sometimes even non-Yolngu Aboriginal people from places as far as the Kimberley, or mixed groups like Sydney-based Bangarra Dance Theatre and the national Indigenous dance academy (NAISDA). The *bunggul* and other public displays of Yolngu cultural knowledge at Garma is also a way to generate interest in traditional knowledge for young Yolngu, to renew the *garma* ceremonies that might be less frequently practiced, and to show them that *Balanda* value this precious cultural knowledge which has demonstrable relevance to modern Yolngu livelihoods. Since 2004 the Garma *bunggul* has included a substantial cash prize for the clan group considered the best performers, which serves as an extra incentive to widely-dispersed Yolngu to mobilise for *bunggul* at Garma. The motivation behind *bunggul* however, is something much more significant than a 'prize'. Franca Tamisari (2000: 151–152) makes the point that Yolngu *bunggul* is art, law and an act of love. She writes,

> Dancing in any Yolngu ceremony ... (is) an event in which knowledge associated with country is transferred, judged, asserted, and negotiated, and through which obligations are fulfilled by offering help to, and demonstrating love and compassion towards one's relatives ... 'Yolngu dance because they hold the Law'.

Bunggul is the space in which Yolngu epistemological difference is made visible to Garma visitors, with multiple levels of meaning available to differently educated viewers. The humorous and spectacular dancing of the professional Red Flag dance group provides an account of cultural contact with the Macassan fishermen with whom they traded goods, words, names and kinship for as much as 700 years (Macknight, 1976). Their performance is both as an historical account, a claim to the capacity for historicity (Rosaldo, 1989) but more importantly is a continuing claim for the recognition of Yolngu sovereignty and their capacity to conduct international diplomacy and trade on their own terms over an extended period of time. The retelling of the story of the Macassan trade has become somewhat idealised, but it is a message for other Australians about the persistence of Yolngu forms of sovereignty and an illustrative ethical model and demand for a culturally and materially deeper reciprocal relationship from still-colonising Settler-Australia.

Garma is a deep pedagogical exercise, both for young Yolngu, and for *Balanda* who in most cases have no or very little knowledge of the Yolngu world. Some Garma participants are tertiary students enrolled in courses at the University of Melbourne, RMIT University or Charles Darwin University, but Garma effectively makes students of all its visitors. Visitors learn to sit quietly with Yolngu women and watch them weave, or go bush to learn how to make a spear with the men. These visitors learn a simple new skill through watching and doing, but more significantly they learn about a Yolngu way of learning and teaching; a profound exercise in encountering Yolngu difference at work that commonly results in tears of frustration and the exhilaration of hard-won understanding. Seeing corporate heavyweights and national bureaucrats become playful enthusiasts for their gender-respective baskets or spears, attentively following their Yolngu instructors, will not immediately change the raw politics of land ownership and mineral rights in this country, but it does open the possibility of a deeper understanding of ontological difference and dialogue that might make such changes possible. The themes of Mandawuy Yunupingu's music and writing capture this spirit, as he explains with Howard Morphy (2000: 494),

> The (Yothu Yindi) band takes on the same agenda to what I did in teaching really. But I'm a musician instead of a teacher. Our objective is to bring about a balance and understanding – a true sense of equality... it's the difference we want to maintain, not the sameness. The sameness can be classified as assimilation. That's what we don't want – we don't want to be assimilated – to think like a white man.

The metaphor (and practice) of Yothu Yindi's music exemplifies this philosophical-ethical approach to actively engage Settler-Australian and wider, global modernity. Their music communicates and reflects the complex experiences and aspirations of its Indigenous audiences, and also draws non-Indigenous listeners into that affective life-world. Dunbar-Hall and Gibson (2004) demonstrate that Aboriginal people have been finding spaces to communicate their experiences and make a living in the Australian music and entertainment industry over generations. The Yothu Yindi rock band sit firmly within this tradition, while having a specific, local and regional experience, particularly the emphasis of using (Indigenous) *language* and naming *country* in songs (Dunbar-Hall & Gibson, 2004: 192– 211). They have combined the technology and musical idioms of western rock, reggae, country and western and techno dance music with the lyrics and musical styles from popular (and in some cases revived) Yolngu *manikay* (song traditions). This innovation on existing musical traditions and the dynamic Yolngu musical context is explored in detail by Aaron Corn (2009) in his discussions with Mandawuy Yunupingu. This virtuoso cultural hybridism is both a creative and generous act of sharing from confident people, secure in their identity, language, land and culture; and at the same time the urgent strategic manoeuvre of an otherwise culturally besieged people suffering intergenerational crisis.

Since the arrival of the massive mine and its town of 5000 residents, the corrosive effects of alcohol and social dislocation have worked their way through three generations of Yolngu living in the Miwatj ('sunrise'- northeast Arnhem Land) region. Yolngu communities gathered into the old mission settlements of Yirrkala or Gunyangara, next to the mine tailings mountain, are suffering the extreme physical and mental health afflictions of other colonised peoples in Australia and elsewhere, with levels of violence and premature death that create a perpetual air of mourning and crisis. As one Yolngu family member from Yirrkala said of the experience of camping at the Garma Festival, 'this is the best it gets all year: there's no drunks, there's plenty of good food and there's *bunggul* and the kids' performances making people proud!'

Garma as Yolngu Philosophy

As well as being the name of a 'festival,' Garma is a concept and practice of the Yolngu people of north-east Arnhem Land. Discussions of Garma need to distinguish between uncapitalised *garma* as a form of Yolngu public ritual religious knowledge and practices associated with funery rites, and capitalised Garma, the event. As Yothu Yindi Foundation background notes explain (1999) for a *garma* to take place as a Yolngu public ritual, there has to be a negotiation between competing, sometimes structurally hostile, but interdependent groups. When this resolution has been made, a spear is thrust into the earth and the *garma* ceremony can proceed around that point. Similarly, for Garma (the festival) to involve the Yolngu clans who form the heart of the event by performing the evening *bunggul* (dance/ceremony) requires complex inter-clan political negotiations on a number of levels, from the sacred ritual and religious to the economic. The central point for Garma is the ceremonially painted *larrakitj* (upright log coffin) installed at the centre of the *bunggul* ground. At another level Garma also calls upon the non-Yolngu guests to enter relations of reciprocity and negotiation with their Yolngu hosts whose land they are on. This reciprocity includes showing respect for Indigenous protocols and opening to Yolngu epistemologies, including the importance of various spirits and spirit-beings to this place and the Yolngu world. Describing the significance of the site and its relationship to the spirit-being *Ganbulapula*, Gaykamangu *et al*. (1999) explain,

> At Gulkula, he formed an open area, called yati, or a garma, for public ceremonials, for all the different Yirritja clans. And they gathered there together over the years, for ceremonies, especially for Yirritja mortuary ceremonies, where the bones of the deceased would be crushed and placed in hollow log coffin, and their spirits would be sent with a sacred string into the spirit world.

> Even today, the Gumatj owners continue to call people together with the spiritual yidaki across the nation and the world, to come together in the spirit of garma. Using the old Yolngu ideas, the modern day spirits which come are exposed to a modern garma, where they come together to learn, to share and to develop ideas and celebrate together through art, through dancing, through radio, television, computers, internet, learning yidaki, learning about medicine, law, many different themes worked together.

In this explanation we are all 'modern day spirits' called together at Garma by the sacred *yidaki* (didjeridu) to learn. This learning is offered very much in the mode of Smith's *Decolonizing Methodologies* (1999), following and respecting Indigenous protocols directed by community elders in an historical context where that authority has been (and continues to be) undermined by processes of colonial domination. Garma offers settler Australia opportunity for deep intercultural dialogue, and a model for how the national story might be constituted differently through a shared process of decolonisation (Rose, 2004). Extending the concept of *garma* to Yolngu- settler interaction invites participants from the dominating settler culture to consider not only their over-determined *difference* of cultural interests from Yolngu, but to recognise their *interdependence* with them and the land, culture and life world that they maintain. Mandawuy Yunupingu (Yunupingu & Morphy, 2000) has articulated these Yolngu ideas of radical interdependence and balance in his music and writing. The collected artist statements in the remarkable *Saltwater* (Buku-Larrngay Mulka Centre, 1999) publication are an authoritative statement of the inter-relatedness of human beings and the natural and spiritual worlds.

Yolngu intellectual and political leaders have highlighted their rich ritual, artistic and intellectual traditions in an attempt to reconcile with colonial modernity. Closed off historically from the option of a nationalist anti-colonial struggle, the cultural-political expressions of Yolngu through Garma and other cross-cultural spaces (such as the dynamic visual arts movement) seeks to elicit and work with emergent strands of decolonising Australian nationalism. Yolngu intellectuals have articulated this through Indigenous metaphors and systems for balancing dichotomous tensions. The Yolngu and broader Indigenous Australian struggle is almost entirely framed as being for the recognition of both cultural difference *and* full citizenship entitlements (overcoming disadvantage) within the Australian nation, and in the process transforming it.

Conclusion

Mandawuy Yunupingu (2001) says of one of Garma's purposes,

> We're living in fluid times, trying to discover in more profound ways what it is to be Australian. I think the vast majority of Australians would agree that Aboriginal Australians have a special contribution to make to that. But there seems to be a problem. I think most non-Aboriginal Australians accept that there is a deep intellectual strength to Aboriginal knowledge, but they seem to think of it as a mystery. I hope we are less of a mystery now.

Indigenous cultural festivals are a powerful medium for cross-cultural contact that can displace and reframe those 'mystifying' characterisations as deeper understanding through personal, embodied experience. While dominating cultures easily fall into habitual stereotyping, state management (as Henry, 2008 argues) or New Age romantic misapprehensions of Indigenous cultures; cultural festivals at very least provide opportunities for direct encounters with Indigenous people that can counteract some of these routine colonising practices. Risks remain, however, in emphasising festivals and Indigenous cultural tourism and cultural marketing more generally as a cultural and livelihood strategy. Packaging cultural practices as a commodity for consumption changes established relationships and identities. The Comaroffs (2009) point out that reframing culture as a commodity within the circuits of neo-liberal capitalism has some peculiar effects from the 'corporatisation' of tribes to the production of an 'identity economy'. In the context of the limited choices available in the Australian context, Yolngu are clearly choosing ethno-commodification as one strategy for engaging modernity, over the option of cultural assimilation pushed by the previous conservative Coalition government and passively pursued by the current Labor one.

Cultural festivals provide a potent space for intercultural accommodations to be negotiated on largely Indigenous terrain, strengthening Indigenous agency and resetting the terms of cross-cultural engagement for at least the duration of these staged encounters. Cross-cultural performances have long been a part of the repertoire of strategies of Indigenous cultural survival and assertion, sometimes even in contexts where those performances are part of the colonial exploitation of culture. Intensified globalisation shifts the terms of this engagement. Indigenous cultural activism is moving beyond an emphasis on contesting the colonising-national story's exclusion of Indigenous peoples and identities, to engaging with an emergent global sphere, which simultaneously reinforces specifically local identities and forms of governance. This is not happening with the same intensity everywhere, and it is certainly not a claim for a homogenising globalism, however cultural performances and celebrations are assertions of Indigenous power in this shifting context.

This chapter has tried to understand Garma as a festival and broader cultural event through the kaleidoscopic prism of Indigenous cultural politics. Indigenous peoples have deployed many strategies for resisting social-Darwinist assumptions of their 'disappearance'; not just through violence and direct engagement with state politics, nor just Scott's (1985) 'passively' resistant 'weapons of the weak', but also with the remarkably generous and insistent gifts of cultural life. Throughout the history of contact with cultures of domination, Indigenous communities have asserted the vibrancy of their people, their land and their cultural life through sharing the sensual enjoyments of place, music and dance, food, games, work and sexuality, through to the closely connected depths of philosophy and religion (not necessarily separated from these enjoyments as in the dominant Western traditions). These acts of generosity have been both attempts to educate and civilise the dominating cultures into a proper ethics of living, as well as a direct political assertion of various forms of existence and sovereignty through means not recognised by the dominating cultures. This aspect of Indigenous cultural assertion has been generally misunderstood and under-theorised through the lens of either romanticism or 'salvage anthropology' as cultural revival and survival, rather than as a seriously political and ethical practice of immersed, embodied experience. These celebrations are serious, joyful and urgent acts of cultural politics.

References

Buku-Larrngay Mulka Centre (1999) *Saltwater: Yirrkala Bark Paintings of Sea Country*. Neutral Bay, N.S.W.: Buku-Larrngay Mulka Centre in association with Jennifer Isaacs Publishing.

Chandler, J. (2007) Whose coup? Canberra and clan both celebrate a deal. On WWW at http://www.theage.com.au/news/national/whose-coup-canberra-and-clan-both-celebrate-a-deal/2007/09/21/1189881777542.html?page=fullpage#contentSwap1. Accessed 26.06.10.

Christie, M. and Greatorex, J. (2006) *Yolngu Life in the Northern Territory of Australia: The Significance of Community and Social Capital*. Darwin: Charles Darwin University. On WWW at http://www.cdu.edu.au/centres/inc/pdf/Yolngulife.pdf. Accessed 26.06.10.

Commonwealth of Australia (2007) *Northern Territory Emergency Intervention Act*.

Comaroff, J.L. and Comaroff, J. (2009) *Ethnicity, Inc.* Chicago: University of Chicago Press.

Corn, A. (2009) *Reflections & Voices: Exploring the Music of Yothu Yindi with Mandawuy Yunupingu*. Sydney: Sydney University Press.

Dunbar-Hall, P. and Gibson, C. (2004) *Deadly Sounds, Deadly Places: Contemporary Aboriginal Music in Australia*. Sydney: UNSW Press.

Gaykamangu, W., Marika, R. and Christie, M. (1999) Gulkula and Ganbulapula. On WWW at http://www.garma.telstra.com/1999/gbackground.htm. Accessed 26.06.10.

Henry, R. (2008) Engaging with history by performing tradition: The poetic politics of Indigenous Australian festivals. In J. Kapferer (ed.) *The State and the Arts: Articulating Power and Subversion* (pp. 52–69). New York: Berghahn Books.

Macknight, C. (1976) *The Voyage to Marege: Macassan Trepangers in Northern Australia*. Carlton, Vic.: Melbourne University Press.

McMillan, A. (1999) Report on the first Garma, Yothu Yindi Foundation. On WWW at http://www.garma.telstra.com/1999/workshop.html. Accessed 26.06.10.

Rosaldo, R. (1989) *Culture & Truth: The Remaking of Social Analysis*. Boston: Beacon Press.

Rose, D.B. (2004) *Reports from a Wild Country: Ethics for Decolonisation*. Sydney: University of New South Wales Press.

Scott, J.C. (1985) *Weapons of the Weak: Everyday Forms of Peasant Resistance*. New Haven, CT: Yale University Press.

Smith, L.T. (1999) *Decolonizing Methodologies: Research and Indigenous Peoples*. London: Zed Books.

Tamisari, F. (2000) Knowing the country, holding the law: Yolngu dance performance in north-eastern Arnhem Land. In S. Kleinert and M. Neale (eds) *The Oxford Companion to Aboriginal Art and Culture* (pp. 146–152). Melbourne: Oxford University Press.

Williams, N. (1986) *The Yolngu and their Land: A System of Land Tenure and the Fight for its Recognition*. Stanford, CA: Stanford University Press.

Yothu Yindi Foundation (1999) Garma background notes. Darwin: YYF. On WWW at http://www.garma.telstra.com/1999/gbackground.htm. Accessed 26.06.10.

Yothu Yindi Foundation (2002) *Garma Festival*. Promotional DVD. Darwin: YYF.

Yunupingu, M. (2001) What Is Garma? Yothu Yindi Foundation. On WWW at http:/www./garma.telstra.com/aboutgarma.htm. Accessed 26.06.10.

Yunupingu, M. and Morphy, H. (2000) A balance in knowledge: Respecting difference. In S. Kleinert and M. Neale (eds) *The Oxford Companion to Aboriginal Art and Culture* (pp. 493–496). Melbourne: Oxford University Press.

Chapter 8
'Our Spirit Rises from the Ashes': Mapoon Festival and History's Shadow

L. SLATER

Introduction

In 1963, the Queensland police forcibly removed Aboriginal people from Mapoon mission, on Western Cape York, in the far north of Australia, and then burned their houses to the ground to prevent return.

Forty-four years later, on 18th November 2007, the rebuilt town held the inaugural Mapoon Day festival. Old Mapoon, as it is often known, is a small Indigenous township with a population of approximately 200 people on the traditional lands of the Tjungundji people. It is a one-and-a-half hours drive from Weipa – over a red, dusty road that is near impassable in the wet season – where the world's largest bauxite mine operates. It is country rich in minerals, which for some has meant jobs and prosperity, others dislocation and for perhaps more than a few, elements of both. I came to the Mapoon festival accidentally, after sitting next to the Principal of Mapoon Primary School on a local airline flight, who invited me to their festival, where they were hoping Midnight Oil – an iconic Australian rock band known for their stance on Indigenous rights – would play 'Beds are Burning' (1987) (a song about Aboriginal land rights, which many in Mapoon felt referred literally to the burning of their town in 1963). My interest in this event lies not only with what, at least for me, is a fascinating and heroic local history, which tells too much about Australian's race relations, and the too often forgotten violence of capitalism, but also in taking Isabelle Stengers' (2005: 188) advice to, 'take your time to open your imagination and consider this particular occasion'. In this chapter, I discuss

Mapoon (the place), the Mapoon Day festival and my experience encountering people and their stories at this event.

By almost anyone's standards, the Mapoon Day festival was the most modest of events. In the morning, there was a parade (Figure 8.1), locals marched with banners that displayed the newly established town mottoes: 'Our Spirit Rises from the Ashes' and 'Our People, Our Place, Our Future'. They marched to the school grounds where a fete was held, and in the evening, after a few hours of rest, a band played in the sports centre. Indigenous cultural performances, Rosita Henry writes, make people present in a world that has rendered them absent (2000: 587). It is the presence, or presencing, that I am most concerned with here. What is made present by Mapoon festival? In attending to the locality of culture and the specifics of place, as Homi Bhabha (1994: 147) asks, 'what are the forms of life struggling to be represented in that unruly "time" of national culture ...'? What might a humble event such as the Mapoon festival have to illuminate about belonging in and to our unruly time?

At the school grounds, I was introduced to local elders – Susie Madura and Grace McLachlan – who were raised on the mission and, although they were away working at the time of the burning of Mapoon, had vivid

Figure 8.1 Local residents getting ready for the Mapoon festival march (Photo: Lisa Slater)

and visceral memories of the violent destruction of their home. Soon after we met, I told them what brought me to Mapoon and Susie's response was to tell me about a woman who was undertaking research on, what Susie called, 'the dormitory days'. The researcher had approached several of the older residents who had lived in the mission dormitory and asked to hear their memories of those times. Susie adamantly told me, if the woman wants to talk about those days, then she must first hear about the burning of Mapoon. The burning, she told me, was far worse than the dormitory days, and they were bad. Susie's story might be apocryphal, but I took her point and with it the invitation for an in situ history lesson. I had a car and between the day's events we – Susie, Grace and myself – drove around Mapoon visiting sites of significance for the women (Figure 8.2).

Two years later, Grace and I were having afternoon tea at the Cape Café, Weipa. Reflecting upon the dormitory days, she told me 'we said our prayers and they locked the doors'. If parents spoke up for their children or family, they risked being sent to Palm Island: little more than a penal colony for Aboriginal 'troublemakers'. Despite this, parents agitated the government to get their children back home, but to little avail until the dormitory nearing burned down. A spark started a fire behind

Figure 8.2 The last remaining house that survived the burning of Mapoon (Photo: Lisa Slater)

the locked doors: thankfully, the children were rescued. Grace said proudly, her father stood his ground: she left the dormitory and moved to an outstation with her family. This story – of violence, defiance and reclaiming a place in the world – was writ large at Mapoon and its festival.

So this is the truth what I've written, and may our God help us to overcome the evil. (Roberts, 1975b: 13)

Cape York has a long history of national and global connections. For thousands of years, there have been Indigenous networks of trade and exchange across the country and with Australia's northern neighbours. In 1605, the Dutch navigator Jansz landed near the site of the modern-day town of Weipa and in 1770 Captain Cook, sailing between the mainland and the Torres Strait Islands, named Cape York Peninsula. But as is well recorded, it was British colonisation that brought the most disruptive and destructive forces to Indigenous people and their sovereign lands. In the 1800s, the cattle industry made incursions into rain-soaked Cape York, and with it came a period of frontier violence; the missions followed, providing 'safe haven' for the traditional owners (Roberts, 1981). In late 1891, a church mission was established on Tjungundji land; initially known as Batavia River mission, it was set up by Moravian missionaries on behalf of the Presbyterian Church and with the financial assistance of the Queensland government. According to historian Geoff Wharton (1996) within a few years the mission became known as Mapoon, which means a 'place where people fight in the sand hills', and as the mission's influence grew it extended into surrounding lands and other traditional owners groups moved or were moved onto the mission. When in 1901 Mapoon was gazetted as an industrial school, children – 'waifs and strays' as Protector of Aborigines, W.E. Roth referred to them – primarily from the Gulf country were removed from their families and sent into the dormitory system at Mapoon (Wharton, 1996).

As in many areas of Queensland, South Sea Islander peoples – from Samoa, the New Hebrides and the Solomon Islands – were 'brought' to Mapoon, in this case, to help the missionaries train Aboriginal people in carpentry, agriculture and cattle industries (Wharton, 1996). Some people stayed, married locals and raised families, and this cosmopolitan history is reflected in the people and place. It is here, or at the industrial school, that I return to Susie and Grace. Both of the women lived in the dormitory, although their families were in Mapoon. On our tour, one of the sights of significance was the ruins of the old school and dormitory. Clearing grass and weeds, Grace exposed the schoolhouse foundations, not much more than a few stray bricks; and we scoured the area for the missionaries' head stones. Despite speaking of the dormitory days as harsh times in which they yearned for their families, they appeared to harbour no anger for their teachers and 'keepers'. Outrage was reserved for the government and police who sanctioned the burning of Mapoon.

The stage had been set early for the removal. In 1897, when the *Aboriginals Protection and Prevention of the Sale of Opium Act* was introduced, missions became increasingly under the influence of government legislation and policy (Wharton, 1996). The mission itself, according to the people of Mapoon, was a harsh place and people were regularly severely punished and expelled for minor infringements and for defending themselves (Roberts, 1975b: 8). In the 1950s, bauxite was found on west Cape York (Roberts, 1981: 10). As early as 1953, the Queensland government and the Presbyterian Church (and some argue Comalco mines) were in discussion about the close of the mission, and although locals were aware of the meetings, they were not consulted (Roberts, 1975a). At a meeting at Mapoon in 1954, a joint decision between the government and the church was made to close the mission and, in the language of the day, people were to be civilised and evacuated to Weipa or other stations, or to 'assimilate those ready for exemption [from the Protection Act] into the Australian way of life' (Wharton, 1996). The people of Mapoon's protests and concerns went unheard, and instead many compelling, although untrue, reasons were advanced for the mission closure: poor agricultural soil, lack of water, poor accessibility for loading and unloading boats and inability for planes to land (Roberts, 1981; Wharton, 1996). Grace recalled the missionaries informing them that they had to move because there was not enough water or it was of a poor quality; only much later was she told that Mapoon is located on fresh springs and would never run out of water.

Grace learned of the burning of Mapoon from the radio. On 15th November 1963, after a joint decision between the Queensland government and the Church, police (including those from Saibai Island, Torres Strait) forcibly removed people from their homes and before all the residents left the mission, Old Mapoon was burned to the ground (Roberts, 1981: 115). Earlier in the day, police began to round up those believed to be troublemakers, and secured them in the mission house over night, before sailing for Red Island Point, now Seisia, on Cape York Peninsula. 'The Church, cookhouse, school, work shop, butcher shop, store all burnt down', Simon Peter recalls, 'They left the medical store but they took away all the medicines' (Roberts, 1975b: 14). Many people reported becoming sick and some believed people died because 'burning down a home means you burn the whole body of our aboriginal custom and they die so fast'

(Roberts, 1975b: 14). Most people were relocated to the northern peninsula of Cape York, to Bamaga and an outlying area, Hidden Valley, renamed by the government as New Mapoon. In the weeks prior, Grace had been listening to radio reports that the Queensland (QLD) government would 'effect the transfer of families' from Mapoon, but she did not believe it. She remembers her family ringing with the news and telling her they were 'sitting around the wireless crying'. Susie's mother was in hospital in Thursday Island and did not have the opportunity to return home to collect anything. A final group of people stayed on for a few months after the burning, but lack of food and medicine forced them to leave (Roberts, 1975b: 14). Susie blamed the 'devil policeman Hughes', the officer in charge who brought the Saibai police, for the burning. For Susie and Grace, the pain of betrayal was still raw.

On several occasions I became aware that the sites we were visiting were memorials, which evoked nostalgia and grief from the women. I stood, no doubt awkwardly, alongside them, in what felt like the place of potentially very different lives; and when they hurried back to the car I took their cue and scurried after them. It is strange how objects as prosaic as a brick, a lonely pylon or a memory of a pier demand and receive the reverence of a stranger. One of the most moving sites was at Point Cullen where the lugger, which carried people and goods between Bamaga and Thursday Island, used to moor. On the beach, my gaze followed Grace's, along her aged arm, as she looked out into the past for the boat that took her as a girl to work as a domestic servant for Thursday Island pearlers. I asked Grace if it was the mission's decision to send young people to work on the Islands or to Cape York cattle stations: she said, no, it was that of the Director of Aboriginal Affairs. According to her, the Director would tell the mission that a certain number of girls were needed somewhere and they would be promptly put on a sailing lugger and sent north. They preferred mainland girls, she said, because they had better English. It was not until she got married that she became aware of just how young she was when she started work. When her fiancé applied for her birth certificate, he was shocked that she was still so young. The missionaries, she told me, must have lied about her age. It took three days to get to Thursday Island. The pittance that they received in payment indicates that they were sent into exile and servitude.

After their removal, the traditional owners continued to lobby the government for the reopening of Mapoon, but to no avail. Despite this, in 1974, about six families returned to their country to rebuild their destroyed community (Roberts, 1975b; Wharton, 1996). Susie was one of those people, and on our tour she took me to the site where they built the first service shed – pointing out the remaining pylon she proudly told me that they had lifted it into place by hand. It was here that they held their first Christmas party, and she said, they sang 'ain't it good to be back home', then went into separate corners and wept.

What does it mean to lose a place and re-inhabit it? It is tempting to make those who re-established Mapoon into heroes. But I am alert to the risk of imposing a singular experience of fortitude and resilience over the vulnerability, pain and confusion they shared; and that affected me. I would also be missing the point of what was being made present to me: a wounded space. Deborah Bird Rose (2004: 34) writes that a wounded space is 'a geographical space that has been torn and fractured by violence and exile, and that is pitted with sites where life has been killed': Australia, not just Mapoon. For the moment, rather than making them into heroes I want to reflect upon what people are able to do, think and feel because they belong. Colonialism attempts to displace and detach people from bodies – people, country, knowledge and law – from which their flourishing is derived, and re-attach them to *strange* bodies (Stengers, 2005: 191). Colonialism says to some *bodies* you do not belong but you can be here with us if you relinquish your sovereignty for ours. People are dispossessed and their systems are manipulated, Stengers continues, so that the coloniser can entertain the notion (or fantasy) of peaceful cohabitation. Bodies are disabled through dispossession, but attachment 'enables people to think, to feel, to invent, to be able or to become able' (Stengers, 2005: 191). Attachment to Mapoon enabled people to return, in the face of such opposition, and rebuild their town: to lift, by hand, the pylons into place. It is not enough to be simply part of society, as Stengers cautions. Obligations to people and place require forms of attachment and belonging. But how can 'we' belong in and to a wounded space? And what does it do to people (and places) when their memories and the stories of violence have largely gone unheard by those who have profited? And, what does it mean to bring these memories to life in another time? These losses occurred in, and are particular to, Mapoon and its people, who have very different relationships to those memories than other Australians (Rose, 2004: 51). And in other forms of dispossession, what stories do not get told or become detached from the life of the people and place? What obligations to history and place then was the Mapoon festival performing?

At one stage, Susie pointedly asked me, why did they remove us? Why did they make us leave when the mining lease does not come onto Mapoon? She could answer this question far better than me. I felt she was asking of both me and my people – white, settler Australians – for a moral engagement with the past in the present. She was holding open history; even affording me the privilege of walking around in it, but with it comes responsibility and appreciation (Rose, 2004: 11). I was being called upon to ethically engage with colonial history. But what does this mean – who or what am I responsible to? Susie and Grace were interrogating and antagonising Australian history and they did not need me or anyone else to tell them of the silences and incoherence of national narratives. Nor that settler Australians have an aversion to getting too close to ongoing pain and loss. Like my own: my inclination to insert those who rebuilt Mapoon into a grand narrative of heroes is telling of my desire to make sense from senseless violence and to buffer myself from vulnerability and loss. Clearly, Susie and Grace made their mark on me: I am not only fond of them, but feel indebtedness and gratitude as well. Without them, their history lesson, copious cups of tea and conviviality, my imagination might not have been as open to Mapoon and the festival. Surely openness to the past in the present also requires attentiveness to the present.

Next to Susie's house is one of the few remaining structures that survived the burning. It was her brother's house, now abandoned – one room, on stilts, made from rippled iron sheeting – rust stained but with a youthful air (Figure 8.2). On the way to Point Cullen, she pointed out the little house: I was mesmerised. Back at her place, we walked around it; when they went inside I lingered. I was drawn to the little house; felt its visceral longing for a connection with the living. Reluctantly, I followed Susie and Gracie inside. Later, I abandoned collecting mangoes to return to the little house. I was perplexed; there was no sense that it held memories of violence. Standing before the little house, I felt it asking me to forget, to just be with it for a few minutes.

From the Ashes

The smallness of Mapoon festival defies interpretation. It could even be said to mock bolder critiques, such as Victor Turner's (see Kleinert, 1999: 346–347), who writes that the significance of staged performances of cultural identity is that they occur in a liminal zone, a transitional stage 'betwixt and between' the everyday and ritual life, and are a form of social and symbolic action: a critique of the social and historical. Such observations seem much more appropriate for the major Indigenous festivals and cultural performances that have emerged over recent years, including Garma (see Chapter 7), Yeperenye, Dreaming, Barunga and Laura Festivals and the Coming of the Light. They are much more readily amenable to the ideas that Indigenous festivals have become a means of entering into intercultural dialogue, a testimony to ongoing political struggles and, for both Indigenous performers and their audience, provide an important context for the contemporary negotiation and transmission of Indigenous peoples' identities (Kleinert, 1999: 345; Slater, 2007). Like non-Indigenous festivals, they are deployed as a means to enhance community creativity, belonging and well-being and thus nourish community resilience. Notably, as Duffy and Waitt (see Chapter 3) suggest, because festivals are structured events, they bring groups and communities together to mark out particular socio-political, historical and cultural affiliations. Festivals are important events that provide both material and symbolic means of responding to and coping with change (see Chapters 1 and 9). Mapoon festival was a modest and quiet event but like many community celebrations it was not unambitious.

Mapoon is known as a place that was razed to the ground, the festival organisers lamented, but there is little consideration of its triumphs. In 2007, Lisa O'Malley, the School Principal and Tom Corrie, the Mapoon Council Deputy CEO, were community representatives at Cape York Institute Leadership workshop, which required them to initiate and run a community development program. Mapoon Council had applied for a gazetted community day – a public holiday – and it was awarded for 19th November. Lisa and Tom did not want the day to go unmarked, as is most often the case, deciding instead to mark it with a festival. They believed that Mapoon had a lot to be proud of, which needed to be recognised by locals, the broader Cape York communities and Australia alike. They have few of the social problems of other Cape York communities, but their reputation suffers through association. A sad irony was that the positive social situation meant that Mapoon received very little attention and missed out on services and programs designed to ameliorate social disadvantage. There was no internet service, and it was thought that because the school had high attendance rates and reached state averages they would be exempt from government initiatives such as 'pool for school', which saw public pools being built in some Aboriginal townships (although to date, not in the Cape) to improve school attendance. The organisers had great and very ordinary, yet important, aspirations for the festival: they wanted to put Mapoon on the national map and the kids needed activities. Midnight Oil was invited, as was the Indigenous Australian, and former Olympic athlete, Cathy Freeman; a charmingly bold refusal to concede to the status of marginality. So were local organisations and the very active Uniting Church. Notably, Lisa and Tom saw it as an opportunity to renew local identity. After being betrayed and destroyed by the government, Mapoon was re-established, a successful community has grown and development continues. Yet prior to the festival, Lisa said, there was little consideration of how proud they should be of contemporary Mapoon.

The town parade was due to start at 9 am from the Home and Community Care Centre to march to the primary school grounds where the day's events were to be held. Mapoon township consists of not much more than a few streets: many of the homes are tucked in among the bushland that stretches along the beach. When the town was rebuilt the locals had a significant input into the planning, opting for large house blocks to create space and privacy. It was Saturday morning and, as in most places, things were slow to start. While we wait, I asked the Council Chairman what the population was: 'Today, 208', he answered wryly. People began to gather and unfurl large banners in readiness for the march. In discussions and consultation leading up to the festival, the organisers asked people what they wanted as the town mottoes. Consensus was reached: 'Our Spirit Rises from the Ashes' and 'Our People, Our Place, Our Future'. Despite the small number on the march – 30 or so – it remains a powerful symbolic gesture to perform for one's own community, 'our people', not for visiting spectators from government or other altruistic agencies. The festival, and especially the parade, brought a public into being that was attending to Gauguin's famous questions: who are 'we' now, where have 'we' come from and where are 'we' going: a public not dominated by, or responsive to, government agendas. Community festivals create an environment for reflecting upon community values and the narration of collective identity (Duffy, 2005). However, the significance of the festival was not the creation of a collective narrative, but rather that it acted as a space-clearing gesture, which rekindled attachments to contemporary Mapoon – the *now*. The short walk to the School felt both sombre and festive: a rebirth requires a claiming of the present and future, and relinquishing something of the past.

How are lost places re-inhabited? The removal and subsequent burning of Mapoon was an act of dispossession: the return and rebuilding one of re-attachment (cf. Read, 1996). But belonging or attachment (and attendant obligations) is an ongoing process, performed everyday, and re-enacted materially and symbolically. Places, and belonging, must be kept alive by renewing connections, which requires, among other things, invention (Stengers, 2005). In my conversations with locals, there appeared to be intergenerational differences about the meaning of the festival. For the old people the event evoked sadness and they wanted to reminisce. The younger generations saw the festival as about rebirth and leaving the sadness behind. Cultural performances, as Kleinert observes, following Turner, are not simply expressions of social systems, but rather they are also reflective: implicitly or explicitly commenting on social life and the way society deals with its own history (1999: 347). To make a place inhabitable, it must be reinvested with meaning and memories that keep it alive. To do so requires revisiting history, bringing stories of that place to life and also being available to the present and what is emerging. The choice of a festival, over alternative community development initiatives, tells a lot about what is understood as pressing issues in Mapoon. How to generate collective pride in the community as it is now? What to make of its history? This leads me to a further question: how to revitalise belonging to a peaceful place with a violent history from a political time of neocolonialism?

There is a lot of history to live in and with at Mapoon. Contemporary public discourse about Aboriginal communities in general is overwhelmingly negative, to which state and federal governments have responded with interventionist policies and programs to 'close the gap' between Indigenous and mainstream populations health, socio-economic and education disparities (see Altman & Hinkson, 2007; Anderson et al., 2007; Pearson, 2007). After years of neglect there is a lot to do. Even if Mapoon gets very little attention, as already noted, the town's reputation suffers by association (the case for most Indigenous communities), especially with nearby Aurukun. As would be expected of such a tiny town, not a lot happens and, as one local told me, without an event like the festival there is very little for people to do. If not creatively attended to, the subtler and somewhat more ephemeral needs of revitalising connections to people and places, which are fundamental to well-being, could get lost among the din of public discourse and government policies and the brooding past – a violent, yet heroic history that largely remains untold beyond the bauxite deposits. In many Indigenous townships and communities there is much need for improvements in health, housing, education and employment, and so is there for gatherings, not dominated by the busyness of governments, which afford locals the space to revitalise the symbolic and communal life of 'Our People, Our Place, Our Future'.

At one end of the sports centre, kids play basketball, at the other the band, Walker Brothers and Cole. A young man, 19 years or so, takes the dance floor with bravado. A lone figure: his audiences are wallflowers seemingly grateful they will never be asked. His attire is that of a northern cowboy: tight blue jeans, big, shiny buckle, pearl-buttoned shirt and Stetson hat pulled low. His dance moves are contemporary freestyle interspersed with a bit of 'shake-a-leg'. The young man, at least to me, is impressive: playful and cheeky. Early in the evening, Susie stopped him to say she had not seen him for a long time and ask where he had been. 'Been a tourist', he laughingly replied, and moved on. The rain, coming off an east-coast cyclone, is so heavy that it is drowning out the band, as if they have been put on mute.

The festival made present the magic or power of the Mapoon: the strength of spirit, sovereignty and inventiveness to remake 'our place' from what is *here*. 'Our People, Our Place, Our Future' and 'Our Spirit Rises from the Ashes' are assertions of historical continuity, social legitimacy, autonomy and sovereignty. Notably, their collective spirit rises from the ashes but is not of the ashes: the burning was a sad and violent historical event, but it is not the people, place or future. The festival, and the dialogues it initiated, exposed differing intergenerational relationships to history but I would also argue that it does, what Rose (2004: 24) calls, recuperative work. Central to her argument is that recuperative work does not imagine a former time or space of wholeness to return to or a fantasy of a perfect future of completeness, rather it is dialogical and committed to decolonisation. The practice of decolonisation is not about making the world whole again, but by gathering and reassembling the diversity of life, learning to live in and among brokenness. 'Our Spirit Rises from the Ashes' is, to borrow Rose's words (2004: 8), 'twisting violence back into flourishing and life affirming relationships'. The festival created a gentle, quiet socio-cultural space that enabled people to both remember and forget the burning of Mapoon. To reassemble and tend belonging: open to the past, the present and the future.

Acknowledgements

I am indebted to Susie Madura and Grace McLachlan for so generously sharing their stories and time with me. I also thank Lisa O'Malley for the invitation to the festival and interest in my research, and Anmol Vellani for an insightful conversation.

References

Altman, J. and Hinkson, M. (eds) (2007) *Coercive Reconciliation: Stabilise, Normalise, Exit Aboriginal Australia*. North Carlton: Arena Publications Association.
Anderson, I., Baum, F. and Bentley, M. (eds) (2007) *Beyond Bandaids: Exploring the Underlying Social Determinants of Aboriginal Health*. Papers from the Social Determinants of Aboriginal Health Workshop, *Adelaide, July, 2004*. Darwin: Darwin Cooperative Research Centre for Aboriginal Health.
Bhabha, H. (1994) *The Location of Culture*. London: Routledge.
Duffy, M. (2005) Performing identity within the multicultural framework. *Social and Cultural Geography* 6 (5), 677–692.
Henry, R. (2000) Festivals. In S. Kleinert and M. Neale (eds) *The Oxford Companion to Aboriginal Art and Culture* (pp. 586–587). Oxford: Oxford University.
Kleinert, S. (1999) An Aboriginal Moomba: Remaking history. *Continuum* 13 (3), 345–357.

Pearson, N. (2007) *From Hand Out to Hand Up: Cape York Welfare Reform Project*. Cairns: Cape York Institute.
Read, P. (1996) *Returning to Nothing: The Meaning of Lost Places*. Melbourne: Cambridge University Press.
Roberts, J. (1981) *Massacres to Mining: The Colonisation of Aboriginal Australia*. Melbourne: Dove Communications.
Roberts, J.P. (1975a) *The Mapoon Story According to the Invaders*. Fitzroy, Vic.: International Development Action.
Roberts, J.P. (ed.) (1975b) *The Mapoon Story by the Mapoon People*. Fitzroy, Vic.: International Development Action.
Rose, D. (2004) *Reports from the Wild Country: Ethics for Decolonisation*. Sydney: UNSW Press.
Slater, L. (2007) My Island Home: Indigenous festivals and archipelago Australia. *Continuum: Journal of Media and Cultural Studies* 21 (4), 571–581.
Stengers, I. (2005) Introductory notes on an ecology of practices. *Cultural Studies Review* 11 (1), 183–196.
Wharton, G. (1996) Mapoon History. On WWW at http://www.mapoon.com/37.html. Accessed 07.06.10.

Chapter 9
Birthday Parties and Flower Shows, Musters and Multiculturalism: Festivals in Post-War Gympie

R. EDWARDS

Recently, much debate has occurred about rural and regional areas and how they have responded to shifts in social organisation, often focusing heavily on economic and demographic changes (e.g. Cocklin & Dibden, 2005). This chapter employs a cultural history perspective in analysing such changes, tracing the historical development of festivals in Gympie, a town in southeast Queensland with a population of approximately 15,000.

Gympie, founded after gold was discovered there in 1867, has a long history of festivals. It hosted the first Queensland Eisteddfod in 1885, and holds a successful agricultural show dating back to 1877. Seasonal flower shows began in 1880, run by the local show society. In 1900, the recently formed horticultural society began holding competitive monthly flower shows, awarding prizes for best seasonal blooms, best arrangements, table settings and so on. In 1917, the Gympie Birthday Celebrations began, commemorating Gympie's historic gold find. It was held again, in 1927, before becoming an annual event in 1949. In 1982, the Country Music Muster appeared. From humble beginnings, it has grown into one of Australia's biggest country music festivals.

This chapter analyses selected events, demonstrating shifts in the organisational structure, intent and cultural meaning of festivals from the post-war period to the mid-2000s. Festivals discussed are the Gympie Birthday Celebrations, Gympie Gold Rush Festival, flower shows and National Country Music Muster. Changes in festival management and significance mirror the development of the town, its community and values, from local, conservative norms of loyalty and pride in colonial achievements to nostalgic nationalism in the 1980s and more generic, global themes in the 2000s.

The Gympie Birthday Celebrations 1949-2001

The Gympie Birthday Celebrations commemorated Gympie's founding moment, or 'birthday' – 16 October 1867 – the date Englishman James Nash's gold find was proclaimed in Maryborough. The first edition of this annual festival occurred in 1949, on and around that date. Festivities included a carnival in the park, sporting contests, a morning tea for Gympie's 'pioneers' and a cavalcade of service vehicles, marching bands, Boy Scouts, Girl Guides and floats made by community groups.

This annual festival was fuelled by growing interest in Gympie's history and renewed optimism in Australia immediately following the end of the Second World War. The Snowy Mountains Hydro-electric Scheme epitomised this post-war optimism nationally (Day, 2005), while in Gympie a new water treatment plant, the Wide Bay Dairy Cooperative Factory and a planned Nestle factory were symbols of this hope for the future. An annual celebration of the town's progress from frontier gold rush town to civilised, growing municipality was one way of fostering this hope, and if surrounded by the rhetoric of community spirit and town pride, could provide a platform for specific political and social goals (Darian-Smith, 2002: 96).

The Gympie City Council, *Gympie Times* and Gympie Traders' Association organised the festival with local service clubs, most notably Gympie Rotary Club. This constellation of organisers focused on civic aspects, carefully constructing the festival to foster pride among residents. The civic was emphasised in two important ways: first, Gympie's pioneers (especially James Nash) were honoured, exalting their labour and the progress they represented. In 1953, a monument to Nash was erected where he first struck gold. Thereafter, rallies were held there to honour him. This rally around the monument commemorated the founding moment and established a 'site of memory' (Nora, 1996) that assisted in developing community and senses of place. Progress from wilderness to 'civilisation' started there: Gympie's churches, schools and hospitals were built because of Nash. By celebrating this progress, all residents were expected to embrace Gympie and specific values constructed to present a successful and civilised community.

Second, Rotary promoted civic duty as responsible citizenship. If, as Murphy (2000) argues, the 1950s were Australia's middle-class decade, the 'good citizen' push of Gympie Rotary was true to type. Just before the 1951 Gympie Birthday Celebrations, Rotary published an article titled, 'A Week of Inspiration for Gympie Residents' (*Gympie Times*, 1951: 2), intended to excite town pride and interest in the event. Rotary urged residents to join their local progress association and/or service club, attend and 'honour your church in the proper manner' and to 'condemn less and rely less on Government and local authority administration'. City Council endorsed these ideals, praising the service clubs' contributions, reinforcing notions of respect for authority and community involvement. The *Gympie Times* embellished this message, arguing, 'civic achievement' should be commemorated, particularly public organisations, which have 'given valuable service for community well-being ... [and] have demonstrated the abiding quality of good citizenship' (1952: 4).

This pattern of recognising highly valued citizens was expanded in 1953 to include another, equally important group: children. When the Nash monument was unveiled, Gympie Boy Scout Troop provided the official party's guard of honour. The following year, Girl Guides and Boy Scouts led a torchlight procession from the Nash monument to the night's festivities in the park. The children heard about the importance of honouring the past and its heroes, while the 1954 procession could be interpreted as symbolically passing the torch of heroism, duty, sacrifice and the pioneering spirit from Nash to Gympie's young people. The good citizen push was passed down (Figure 9.1).

From 1950 to 1967, the festival was relatively stable. However, in 1967, Gympie's centenary, the traditional name was used interchangeably with the title 'Gold Rush Festival', despite that name only referring to the year-long centenary celebrations. While uncontroversial, it signified a shift in the event's focus. Gympie's local history and culture was repackaged for tourist, not local, consumption; a repackaging aimed at community economic development (Chhabra *et al.*, 2003). Gympie was not unique in developing tourism-focused events during this period. In North America too, the mid-1960s saw new annual festivals emerge to attract tourists, based on crop harvests, foods, historical events and other local features (Janiskee & Drews, 1998).

The Gold Rush Festival was a tourism-oriented event using Gympie's history as a place marketing strategy, stressing the romance and adventure of the goldfields. Manipulating the community's image to attract tourists ensured the event quickly became a commemoration of a ubiquitous nature, a common strategy in promoting regional heritage (De Bres & Davis, 2001). Gympie residents had long believed their history was important, but it needed incorporating into a more universal history to become relevant to tourists. The 19th-century gold rushes conjure up romantic images of shanty towns appearing overnight, fortunes made and squandered in a day, and a rugged democracy that historic recreations, documentaries and theme parks like Ballarat's Sovereign Hill gold mining town in Victoria have made *de rigueur* (Evans, 1991). At Gympie's

Figure 9.1 Gympie's centenary procession (*Gympie Times*, 1967: 4)

festival, the (imagined) iconography of the Californian Gold Rush appeared immediately, including an incongruous 'Buffalo Bill' float, 'complete with frontiersmen and women and Indian maidens' (*Gympie Times*, 1967: 4). Such images increased during the 1980s, when a competition to find the first 'Nuggety Bill' – the man who best embodied the classic gold miner – began.

Changes to the Gold Rush Festival reflected changes in Australian social values. The 1992 festival acknowledged for the first time the Chinese contribution to the region with a performance by 'genuine Chinese lion dancers and a Chinese dance group'. The *Gympie Times* declared this would allow the town to 'pay tribute to the Chinese community and recognise the role played by Chinese pioneers in developing Gympie from the time of the early gold rush days' (*Gympie Times*, 1992a: 20). This was a significant re-imagining of Gympie's history, given prior emphasis on white colonial achievement – a bold move, even within the context of federal government policies promoting multiculturalism (Office of Multicultural Affairs, 1989). The newspaper openly acknowledged the contribution of Chinese *pioneers* to Gympie's development, something the Cooloola Shire's official history (Cooloola Shire Library Service, 2001), centred on Gympie, did not. Alongside the Chinese, a Philippines national song and dance troupe, a German Alpenrosen folk dance troupe, Swiss mountain trumpets and the usual horses and buggies, military marchers, service clubs and scouts all participated in the festival parade that year (*Gympie Times*, 1992b). However, no multicultural scenes were evident in 1993; the festival reverted to type, raising questions of tokenism regarding the 1992 event.

In 2000, the festival hosted another multicultural celebration. Organisers gained Festivals Australia funding for multicultural workshops, including Samba percussion, Macedonian Gypsy brass band performance, South African singing, Angklung bamboo music, Zimbabwean Marimba, Brazilian dance and human values from an Aboriginal perspective under that year's theme, 'Festival of Nations'. Workshop participants joined the procession and performed a concert in Nelson Reserve. The festival sought to develop community skills and foster understandings of Australia's multicultural nature, while promoting Gympie as multicultural and cosmopolitan. However, most workshops had scant relationship to Gympie itself; no Brazilians or Macedonian Gypsies lived there, suggesting it was more about exotic desires or attracting tourists.

However, some 'multicultural' participants were residents of Gympie. Gympie's social composition had changed from almost exclusively Anglo-Australian or British born to one in which 9.4% of the population were born overseas, including 3.3% from countries whose official language is not English (2001 Census). Significant numbers of residents were born in Germany, Holland and the Philippines. At the 2001 festival, the most notable multicultural event was the Filipino community's participation in the parade, especially given the absence of any obvious 'traditional dress'. They were involved on their own terms, not as some stereotypical 'ethnic minority', as was evident with the lederhosen-wearing German participants in 1992.

These innovations suggest that the Gympie Gold Rush Festival has modernised, recognising changes in society over time – although not so far as to acknowledge the theft of Aboriginal land that mining required. Organisers showed an increased understanding of the reality of multiculturalism within the region, modifying the festival to reflect it. By instituting multicultural celebrations alongside more traditional aspects of the event, organisers have sought to position Gympie as a progressive, modern rural centre proud of its (contemporary) diversity.

Flower Shows: The Rise and Fall of a Community Event

The Gympie Agricultural Mining and Pastoral Society held the first seasonal flower show in 1880. From 1895, the Gympie Municipal Horticultural Society (GMH) took over, adding monthly shows in 1900. Flower shows reinforced ideas of civilisation and ownership of nature by domesticating the conquered landscape through flower cultivation and making the house a home (see also Mayes, Chapter 10). They advanced civic values by beautifying the town, encouraging others to follow that example. These smaller events were numerous and significant players on Gympie's cultural landscape up to and during the 1960s. After that time, their importance diminished markedly.

The early shows were not exclusively flower shows, but horticultural shows, which focused both on aesthetic floral displays and practical matters such as vegetable and fruit growing. By the 1960s, flowers dominated, with only small sections for other goods; only the Gympie Show continued displaying fruits and vegetables prominently. Gympie held monthly GMH Society flower shows and nine annual shows – church, school and Red Cross events, which were usually held in autumn or spring. All were well attended, whether held in a church hall or a school. In sheer numbers, they were major cultural events in Gympie during this period.

Since the first horticultural shows were more focused on the science of fruit and vegetable production, many men participated. When monthly shows were instituted, women came to dominate, as emphasis shifted towards flowers. Holmes (2006: 169) identifies this as part of the gendered roles of gardening: 'women's gardening was more involved with creating beauty' than men's, which focused on agricultural production: growing vegetables. However, men did grow flowers – frequently winning prizes for their roses, in particular – but did not participate in the more aesthetic work of flower arrangements. By 1960, horticultural shows were now flower shows of local significance: the *Gympie Times* recorded winners; photographs frequently adorned page one. Winning at GMH Society Flower Shows was not easy: keen competition attracted entries from places such as Nambour and Maryborough, up to 90 km away.

One aspect of early horticultural shows that continued was the educational demonstrations on horticultural science. In March 1962, for example, entertaining slides and a scientific display were provided:

> Interesting coloured slides of gardens, flower shows and floral arrangements by Mr P Jacobson of Maryborough, followed by a useful talk on the proper use of fertilisers by Mr John Skinner, of ACF and Shirleys Fertilisers Pty Ltd., concluded a most successful evening. (*Gympie Times*, 1962: 6)

Such displays and information evenings did not feature at annual flower shows, demonstrating the society's roots as a group concerned with increasing the region's productive capacity.

The main objective of the GMH Society was the advancement of gardening and floral display. Accordingly, judges often gave feedback to exhibitors. For example, the floral work judge at one show felt 'some of the exhibits could perhaps have had a little more attention, and to show what she meant, Mrs Crittall re-arranged several exhibits' (*Gympie Times*, 1961b: 3). The floral work competitions reinforced aesthetics as exclusively a woman's domain. By correcting entrants' mistakes, Mrs Crittall was ensuring that the women achieved the best aesthetic and thereby social results possible. These competitions honed women's flower arrangement skills for dinner parties and other social occasions, matters that were their preserve.

This was demonstrated clearly by the visit of (former Lord Mayor of Brisbane) Sir Reginald and Lady Groom to Gympie in July 1961. Lady Groom was guest speaker at the annual dinner of the Gympie Business and Professional Women's Club (B&PW). Returning president of the B&PW Club, Mrs Notley, a regular participant in flower shows (including as judge), was responsible for flower arrangements and received praise from the *Gympie Times* for the 'beautifully decorated' room and dinner tables (1961a: 2). Lady Groom's speech observed, 'a town should be a "jewel of civilisation"', reinforcing middle-class ideas of community service and women's role in society by praising 'the work of women in provincial centres', including the B&PW and Quota Clubs, and the Country Women's Association. Flower shows encouraged not only the aesthetics of flowers, but also the performance of norms of civilisation appropriate to white society. Mrs Notley, one surmises, was the epitome of this ideal, if the newspaper's praise is any indication.

This ideal was further emphasised by the Red Cross 'Chelsea' Flower Show, particularly the floral picture displays. The floral pictures at the 1963 event are representative:

> 'My Fair Lady' sits beside a barrow-load of hundreds of flowers and baskets of distinctive species. An old-world lamp gives added charm to the scene. The CWA's display is entitled 'Flanders Field', and includes masses of Flanders poppies in deep red hues encircling a field of 'soldiers'. (*Gympie Times*, 1963: 5)

These exhibits demonstrated Empire nostalgia, domesticity and civilisation. 'My Fair Lady' carefully constructed a scene of domestic civility. The old-world lamp and miniaturised garden created associations with

British sensibilities, fostering nostalgia and loyalty to Britain. 'Flanders Field' was heavy in Empire symbolism, a floral depiction of a battlefield in Europe where thousands died for King and Country. Combined, they clearly articulate Empire loyalty: 'My Fair Lady' is what those men in 'Flanders Field' fought for, the domestic bliss of garden and home. Each of these miniatures, harking back to Britain and Empire while exemplifying domesticity and civilised life, intended to imbue Gympie residents with a positive sense of their heritage, their duties as citizens and the lifestyle they should lead (Figure 9.2).

Flower shows were organised by community and church groups whose members were predominantly middle-class professional women or the wives of professional men. In terms of values, flower shows resembled the Gympie Birthday Celebrations, as Gympie Rotary Club's article before the 1951 birthday festival illustrates. One duty of the good citizen, Rotary stated, was to 'beautify your surrounds'; the horticultural society 'set an example in encouraging civic pride in *gardens and lawns*', demonstrating the value of gardening and flower shows to civic pride (*Gympie Times*, 1951: 2, emphasis added). These events were all part of a push to develop better citizens, promote town pride and conservative values, but this world was rapidly changing.

Figure 9.2 Floral picture, Gympie Red Cross 'Chelsea' Flower Show 1992 (Joy Currie Private Collection)

Whereas the birthday celebrations morphed into the Gold Rush Festival, flower shows declined starkly after the 1960s: nearly all churches and schools held flower shows in the 1960s, while the Gympie Red Cross event – the last standing – folded in 2004. Those involved in flower shows in the 1980s and 1990s identified the ageing volunteer base as the major problem. Joy Currie, organiser of the final 'Chelsea' Flower Show, observed,

> [W]e have a lot of different groups helped us, and they were, like ourselves, having trouble with volunteers, a lot of them were getting into their seventies and eighties … . You get tired … at that age and don't feel like putting in the work. (Interview 26 July 2006)

The second main factor in the decline of Gympie's flower shows was social change. Whereas during the 1950s and 1960s, many women did not work outside the home, many entered the workforce during the 1980s, meaning 1960s women had more time for gardening, their church and groups like the Red Cross. Flower shows were run by church women's guilds, women on school P&Cs, or women on Red Cross committees. They included wives of council aldermen and local businessmen. For example, Mrs Kidd, St John's Anglican Women's Guild, was the wife of a Gympie Alderman, while Mrs Leda Madill, Gympie Presbyterian Church Guild, was married to the local Holden car dealer. Generally, these women did not engage in paid work, and so meetings were held in the early afternoon. These meeting times inhibited working women's active membership of the guilds, preventing the renewal of the ageing volunteer base. Instead of changing meeting times to suit, the guilds continued as they always had. By the 1990s, the results were clear: unable to attract new members, Gympie's Catholic Women's League failed to fill executive positions and folded in 2003, while two Presbyterian Women's Guilds merged in 1987 due to 'so many married women being in the workforce' (Head, 1990). Flower shows were often the first casualties as church guild memberships dwindled.

As flower shows disappeared, the more progressive Gympie District Show Society began electing women to its committee, including into traditionally 'male' areas, such as Ring, Stud and Prime Cattle, and Safety sections. This was partly recognition of women's changing role in society and acceptance of the value women add to the show. The show also attracted both men and women in business and agriculture, providing opportunities for networking, social capital development and socialising that flower shows, old fashioned by comparison, had since long ceased to provide.

Country Music Muster 1982-2006

In 1982, a new festival began in Gympie. The first Country Music Muster (henceforth Muster) was held at Thornside, home to the Webb family (of whom three brothers were successful country music entertainers), to commemorate two significant family milestones. As the *Gympie Times* stated:

> [The] Centenary of the Webb Family settlement at Widgee will be celebrated next weekend on a specially prepared site which was part of the original selection ... The Widgee Muster – marking the Family's 100 years on the property and the Webb Bros' 25 years in Country and Western entertainment – is expected to be one of the largest such celebrations held in South East Queensland. (*Gympie Times*, 1982: 15)

These two events, combined with a second Golden Guitar (Australian country music's highest award) in 1982, gave the Webb Brothers a reason to celebrate. Marketing focused on the Webb family's achievements and organisers proudly announced that all proceeds would go to charity, continuing the Webb Brothers' commitment to charitable enterprise dating back to 1964 (Walden, 2004: 137). The brothers recruited Gympie Apex Club to manage festivities, and a successful first Muster provided momentum for a second and third. Eventually, the Muster outgrew Thornside and, with Queensland Government assistance, shifted to Amamoor Creek State Forest Park, approximately 20 minutes drive from Gympie.

As the event's standing on the national festival scene developed during the late 1980s and early 1990s, country music's worldwide popularity grew, and new venues were built. In 1992, the CrowBar was built as an alternative to the main stage. That same year, the Muster added the word 'National' to its name, as new marketing imperatives demanded a stronger base be created. No longer was the Muster a local celebration, but a major festival designed to attract visitors nationwide. Media coverage and advertising summoned a national audience through material in the rural press, *Who Magazine*, *Women's Day*, and local and national newspapers, while *Sunshine Television Network*, a *Seven Network* affiliate, distributed a Muster Special across Australia and New Zealand. Success was confirmed when approximately 47,000 people attended the 1993 Muster and 10,000 camped on site (Apex Club of Gympie, 1994: 6). The Muster had become a major Australian festival.

Seeking national attention, organisers looked beyond country music. In 1992, patrons could choose between 'traditional country music' and country rock, while between 1994 and 1996, the USQ Hyundai Big Band, bush poetry, bluegrass, country blues and harmonica concerts provided added alternatives. By broadening the programme and developing new venues, organisers were positioning the Muster to compete better with other festivals. Competition for the music tourism market developed rapidly during the 1980s and early 1990s (Gibson & Connell, 2005: 215–261), making a predominantly local focus unfeasible given the corporate realities of festival promotions. Diversifying content also insured the festival against fluctuating trends (the country music boom had largely ended by the mid-1990s) and enabled organisers to offer something new to loyal repeat visitors.

Having survived for 25 years, the Muster has a repertoire of stories to explain its nature and origins. These stories have changed over time, reflecting new realities within the increasingly competitive music tourism industry. One notable change has been the evolving portrayal of ideas of 'country' and 'the bush' as commercial realities draw the Muster beyond 'pure' country music to other, hybrid forms. The terms 'country' and 'rural' are used interchangeably here, signifying non-urban and generally agrarian lifestyles and regions – Muster organisers frequently conflate these terms while presenting images of a rural idyll in marketing campaigns. For Tamworth, on the other hand, traditional country music is central to its place marketing strategy (Gibson & Davidson, 2004).

The first Muster was billed as a 'weekend in the bush'. Staging a country music 'Muster' (literally rounding up cattle or sheep) on a cattle property provided a rural theme important to perceptions of the authenticity of country music. Just as jazz appears more 'authentic' performed in New Orleans, country music might also feel more 'real' in the bush, on a cattle property owned by three country music artist brothers (Gibson & Connell, 2005: 138). Organisers 'believed that a C[ountry] M[usic] festival would be more credible and successful if it were held in the bush far from civilisation' (Apex Club Gympie, 1987: 4), while the event's name cleverly played on country music's tendency to portray rural life in a generally nostalgic manner (Connell & Gibson, 2003: 80; see also Peterson & McLaurin, 1992).

The themes from the inaugural Muster evolved: new representations of the bush and Australian nationality developed as organisers realised the Muster's tourism potential and new commercial realities. The Muster's appeal needed to broaden from a local celebration of a pioneering family, and by 1988 the pioneer theme was greatly diminished. The bush was posited as a place of tranquillity, relaxation and camaraderie unattainable elsewhere, in 'the company of thousands of others who are just out for a good time' (Apex Club Gympie, 1988: 6), in contrast to images of the bush as a place of (almost) exclusively male labour and mateship (Ward, 1966: 99–102).

In 1988, the Muster was part of Queensland's official Australian Bicentenary festivities, adding nationalism to advertising material – perhaps seeking to occupy territory fruitfully claimed by Tamworth's Country Music Festival, which is held around Australia Day to add nationalistic sentiment (Gibson & Davidson, 2004: 394). Muster organisers chose an Anglo-Celtic version of Bicentennial patriotism for its advertising campaign, represented by a group of country music performers, a man on horseback, a homestead, and the Australian flag in the background. Indigenous Australians were conspicuously absent, despite their strong connections to country music and the bush (Gibson & Davidson, 2004; Smith, 2005: 94; Walker, 2000: 14–15). Figure 9.3 demonstrates unequivocally the links between the bush, rural life and country music, reinforcing the importance of country life, particularly the value of agrarianism, or country mindedness in the Australian context (Aitken, 1988: 51), to white Australian identity. Further, it portrays Australian country music as the 'voice of the outback, the country, or the bush,' with 'strong national roots and some independence' from its global counterparts (Smith, 2005: 110).

In 1992, the national roots of Australian country music filtered into the Muster's name. Partly a marketing strategy in the battle for dominance with Tamworth, adding the word 'national' also suggested country music was the true expression of Australianness – given country people and their lifestyle have long been privileged in mythical constructions of Australian identity. In 1993, the Muster's Rural Aid charity programme began, helping solidify that link. A country music festival helping country people survive a severe drought was a powerful idea, and that year it raised $100,540 (Apex Club of Gympie, 1994: 10).

Even as Muster organisers engaged in rural charity fundraising, the event's image shifted away from the rural, as country music's star faded internationally. For example, the 1994 Muster programme contained no trace of the iconic rural images that had been the norm in previous years.

This image change suggests corporatisation had affected the kinds of audiences the Muster could or should attract and occurred concurrently with diversification into other forms of music. Despite the country music focus, organisers seem to have decided the event needed to appear less 'rural' to attract new audiences. Indeed, this 'de-ruralising' of country music, as Gibson and Davidson (2004: 398) note, is an attempt to lure new audiences as much as distance the music from the 'hick' or 'redneck' stigma.

Figure 9.3 Advertisement for 1988 Muster (Apex Club Gympie, 1988)

By 1999, organisers felt confident enough to reintroduce agrarian images as part of a montage that reflected the event's new face: exciting music, good times and a rural image. However, the Muster continued its move away from 'pure' country, featuring rocker Jimmy Barnes in 2004, while 2006 saw two non-country performers in Shannon Noll and Pete Murray given top billing. Commercial realities were taking over. The new marketing strategy focused on corporate entities as well as families, promoting corporate entertainment areas and facilities like the wine bar and restaurant, luxury camping and fine dining on site.

The Muster attracts corporate sponsors that reflect its national profile, and the values it espouses. The Muster is marketed as uniquely Australian, in a bush setting that reflects the 'spirit of country Australia' (Apex Club Gympie, 2006), thereby becoming attractive to many national and multinational companies, perhaps none more so than Toyota. Since 1999, Toyota, as major sponsor, has used the Muster to help 'Australianise' its brand image, promoting the 'Toyota Country' brand in brochures and programmes. The 2006 Muster programme featured an advertisement employing Australian vernacular accompanied by a picture of a rural pub with six Toyota vehicles parked out in front: 'Australia's most popular working vehicles. No worries'. Toyota also uses this strategy at Tamworth, where Toyota is not only the major sponsor, but also runs its own festival radio service (Gibson & Davidson, 2004: 395).

The Australian images Toyota and other companies seek to exploit are stereotypical, masculine, white images. Toyota presents the bush as white oriented, reminiscent of the 1988 Muster advertisement. Figure 9.3 presents a bush 'tamed' by white settlers (note the fence), while the Toyota advertisement (Figure 9.4) includes two white men standing on the pub's balcony, one of whom is wearing an Akubra: the iconic Australian bush hat.

However, this mass commercialism has not affected the essentially volunteer nature of the workforce and strong sense of community the Muster develops. Corporate sponsorship enables the event to operate without overhead costs while, in return, sponsors gain national exposure and an opportunity to 'Australianise' their image, as Toyota has, by embracing the 'country' narrative the Muster has developed since 1982.

Over time, the Muster has created narratives that explore key understandings of both the event and region through its marketing campaigns. In developing these narratives, organisers demonstrated an awareness of the shifting sands of audience interest and sentiment. The event reflected generational change by seeking a younger, more contemporary audience through broadening the music base, and by marketing to suit the

Figure 9.4 Toyota Country: Advertisement in Muster Program, 2006

times, whether nostalgic nationalism around the Bicentenary or a corporate look during the mid-1990s. The 'country' only returned to Muster advertising in 1999, but then it was very much corporate country. In these ways, the Muster has sought to remain relevant and retain strong audience numbers.

Conclusion: From Flowers to Toyotas

The evolution of festivals in Gympie demonstrates how generational change can affect their organisational structure, intent and cultural meaning. Often, these changes reflected the town's development, community and values. Festivals in the 1950s and 1960s helped foster civic responsibility, loyalty, support for local development goals and conservative norms. During the 1970s and 1980s, these festivals went separate ways: the birthday celebration modernised, becoming a tourist attraction as the Gympie Gold Rush Festival, while flower shows did not. The fate of flower shows demonstrates clearly the adverse results of not recognising and responding to generational change. The 1980s also saw a new festival appear, based on a country music tradition and the success of the Webb Brothers. From 1982 onwards, the National Country Music Muster became increasingly commercial, meaning that the original local meanings of the Muster were superseded by nationalist, and then more globalised, corporatised forms with less emphasis on locality, due to market realities and changing community perceptions.

Festivals can thus reflect broader societal trends and demonstrate the character and history of a particular locality. Gympie's festivals developed due to existing cultural resources: regional history and a country music tradition. Using local resources creates a sense of authenticity and ownership. Authenticity, or 'perceived authenticity' (Waitt, 2000: 847), is vital to a festival's success, despite exceptions like that of Parkes (Brennan-Horley et al., 2007). As the local cultural and physical landscape was exploited, the events became integral parts of Gympie's cultural fabric, and remained. However, this is no guarantee of success, as Gympie's flower shows demonstrate.

Gympie is a 'festival town' – its profile is raised by three or more annual events held there or associated with it (Edwards, 2008: 206). Festival towns have the potential to develop positive images that may counteract negative stereotypes associated with country towns suffering the rural crisis, or, in Gympie's case, the 'hell town' tag received in the late 1990s from criminologist Paul Wilson (Edwards, 2008: 7). A positive image can increase tourist-related income for local businesses, while income for community groups benefits them and the town markedly, as that income is invested in local projects. Festivals can also attract new residents, which in turn brings income, new energy and, sometimes, new events to town.

In 2006, Gympie held the inaugural Heart of Gold International Film Festival. Gympie's status as a festival town contributed to event's development, with the event's organiser, a recently arrived resident, taking the name, in part, from Gympie's gold mining past, building on the 'brand' the town has developed through its festivals over time. Marketing strategies for the first festival focused on Gympie's proximity to Noosa, a factor contributing to the town's attractiveness to sea changers. Gympie's ability to attract and develop new events over time is a constant demonstration of generational change, as new people and ideas create the conditions for festivals that reflect and mould the perceived values of the town.

References

Aitken, D. (1988) 'Countrymindedness': The spread of an idea. In S.L. Goldberg and F.B. Smith (eds) *Australian Cultural History* (pp. 50–57). Cambridge: Cambridge University Press.

Apex Club of Gympie (1987) *Commonwealth Bank Country Music Muster: A Profile of Australia's Largest Outdoor Country Music Spectacular*. Gympie: Apex Club of Gympie.

Apex Club of Gympie (1988) The Commonwealth Bank Country Music Muster provides a weekend of entertainment for the whole family. In *Muster Update: The Official Newsletter of the Commonwealth Bank Country Music Muster* (August). Gympie: Apex Club of Gympie.

Apex Club of Gympie (1994) A report from the Muster chairman. In *1993 Australian Farmers National Country Music Muster Annual Report*. Gympie: Apex Club of Gympie.

Apex Club of Gympie (1999) Muster Update: The Official Newsletter of the Toyota National Country Music Muster (July/August). Gympie: Apex Club of Gympie.

Apex Club of Gympie (2006) Toyota Muster 25: Official Program, Toyota National Country Music Muster August 22–27, 2006. Gympie: Apex Club of Gympie.

Brennan-Horley, C., Connell, J. and Gibson, C. (2007) The Parkes Elvis Revival Festival: Economic development and contested place identities in rural Australia. *Geographical Research* 45, 71–84.

Chhabra, D., Healy, R. and Sills, E. (2003) Staged authenticity and heritage tourism. *Annals of Tourism Research* 30, 702–719.

Cocklin, C. and Dibden, J. (eds) (2005) *Sustainability and Change in Rural Australia*. Sydney: UNSWP.

Connell, J. and Gibson, C. (2003) *Sound Tracks: Popular Music, Identity and Place*. London: Routledge.

Cooloola Shire Library Service (2001) *Cooloola Shire: …A Golden Past*. Gympie: Cooloola Shire Council.

Currie, Joy, interviewed by Robert Edwards, 26.07.06.

Darian-Smith, K. (2002) Up the country: Histories and communities. *Australian Historical Studies* 118, 90–99.

Day, D. (2005) *Claiming a Continent: A New History of Australia*. Sydney: Harper-Perennial.

De Bres, K. and Davis, J. (2001) Celebrating group and place identity: A case study of a new regional festival. *Tourism Geographies* 3, 326–337.

Edwards, R. (2008) Gympie, "The Town That Saved Queensland": Popular culture and the construction of identity in a rural Queensland Town. PhD Thesis, University of Queensland.

Evans, M. (1991) Historical interpretation at Sovereign Hill. *Australian Historical Studies* 24, 142–152.

Gibson, C. and Davidson, D. (2004) Tamworth, Australia's 'Country Music Capital': Place marketing, rurality, and resident reactions. *Journal of Rural Studies* 20, 387–404.

Gibson, C. and Connell, J. (2005) *Music and Tourism: On the Road Again*. Clevedon: Channel View Publications.

Gympie Times (1951) Week of inspiration for Gympie citizens. *Gympie Times* 4 October, 2.

Gympie Times (1952) Significance of an anniversary. *Gympie Times* 18 October, 4.

Gympie Times (1961a) Lady Groom told B&PW Club: Much to consider in planning a town. *Gympie Times* 1 August, 2.

Gympie Times (1961b) Outstanding exhibits at flower show. *Gympie Times* 26 September, 3.

Gympie Times (1962) Society Show: Judge said pot plants best she has seen. *Gympie Times* 3 April, 6.

Gympie Times (1963) Chelsea Flower Show best to date. *Gympie Times* 10 September, 5.

Gympie Times (1967) Diversity of floats in the centenary procession is indicated by This 'Buffalo Bill' group complete with frontiersmen and women and Indian maidens. *Gympie Times* 17 October, 4.

Gympie Times (1982) Centenary on original site. *Gympie Times* 18 September, 15.

Gympie Times (1992a) Gold rush Mardi Gras. *Gympie Times – Gold Rush Historical Supplement* 15 October, 20.

Gympie Times (1992b) What a party! *Gympie Times* 20 October, 1.

Head, M.J. (1990) *Surface Hill Uniting Church, 1890–1990*. Gympie: Surface Hill Uniting Church.

Holmes, K. (2006) 'Planting hopes with potatoes': Gardens, memory and place making. In M. Lake (ed.) *Memory, Monuments and Museums: The Past in the Present* (pp. 166–181). Melbourne: Melbourne University Press.

Janiskee, R.L. and Drews, P.L. (1998) Rural festivals and community re-imaging. In R.W. Butler, C.M. Hall and J. Jenkins (eds) *Tourism and Recreation in Rural Areas* (pp. 157–175). Brisbane: John Wiley & Sons.

Murphy, J. (2000) *Imagining the Fifties: Private Sentiment and Political Culture in Menzies' Australia*. Sydney: UNSWP.

Nora, P. (1996) General introduction: Between memory and history. In L.D. Kritzman (ed.) *Realms of Memory: Rethinking the French Past* (pp. 1–20). New York: Columbia University Press.

Office of Multicultural Affairs (1989) *National Agenda for Multicultural Australia: Sharing our Future*. Canberra: AGPS.

Peterson, R.A. and McLaurin, M.A. (1992) Introduction: Country music tells stories. In M.A. McLaurin and R.A. Peterson (eds) *You Wrote My Life: Lyrical Themes in Country Music* (pp. 1–14). Camberwell, Vic.: Gordon and Breach Science Publishers.

Smith, G. (2005) *Singing Australian: A History of Folk and Country Music*. North Melbourne: Pluto Press.

Waitt, G. (2000) Consuming heritage: Perceived historical authenticity. *Annals of Tourism Research* 27, 835–862.

Walden, G. (2004) We've got the country: The Webb Brothers and Gympie's Country Music Muster. In P. Hayward and G. Walden (eds) *Roots and Crossovers: Australian Country Music Volume 2* (pp. 125–146). Gympie: AICM Press.

Walker, C. (2000) *Buried Country: The Story of Aboriginal Country Music*. Sydney: Pluto Press.

Ward, R. (1966) *The Australian Legend*. Melbourne: Oxford University Press.

Chapter 10
On Display: Ravensthorpe Wildflower Show and the Assembly of Place

R. MAYES

> *Wildflower Shows*
> *They arrive*
> *jammed in boxes*
> *their tiny flowers*
> *massing colour on*
> *the sorting room floor.*
> *Delicate spider orchids*
> *dangling from icecream buckets*
> *a sundew sitting in a*
> *cracked saucer.*
> *Holding fading specimens*
> *eager pickers ask.*
> *What name is this?*
> *Amateur botanists mumble.*
> *Names change*
> *we'll let you know*
> *next year.*
> Laurel Lamperd, Southern Scribes, 2007.

Each September since 1983 in the rural Shire of Ravensthorpe, Western Australia, volunteers collect samples of up to 700 wildflower species which are then displayed in the Ravensthorpe Senior Citizens Centre from 9.00 am to 4.00 pm daily over a two-week period. This chapter offers an ethnographic interpretation of this enduring annual event focusing on the 25th show held in 2007. The study contributes to understanding the complex and nuanced role of local wildflower shows in shaping and supporting rural senses of place and of community. Importantly, this particular type of festival, and more specifically this local instance, foregrounds a less-remarked aspect of festivals, namely the (re)production and celebration of place-specific knowledge through validations of, and interconnections between, scientific flower classification and emotive experience. This feature, encapsulated in Laurel Lamperd's poem above, invites consideration of the ways in which local place knowledge and the simultaneous (re)production of 'place' are constituted by a complex layering of rational, objective ways of knowing and those which emphasize emotions, aesthetics and memories. This rural wildflower show not only mobilises both the rational and the emotional in 'making sense of the world' for local residents and for tourists, but also offers insights into the production of place as constituted in and through relations between humans and non-human life forms (Cloke & Jones, 2001; Conradson, 2005; see also Chapter 6).

The Ravensthorpe Wildflower Show is part of a vibrant local culture supporting a range of events including, for example, week-long writer-in-residence programs, the Hopetoun Summer Festival, regular dramatic productions, Australia Day breakfasts and art shows (Mayes, 2010a, 2010b). The Shire of Ravensthorpe, with a population of 1900 (ABS, 2007) occupies 13,000 square kilometres incorporating extensive national parks and nature reserves. Broad-acre farming is the consistent principal industry, along with large-scale mining for a brief period in the late 2000s. A small tourism industry is supported by the long-standing volunteer-based Ravensthorpe Visitor Centre and Museum. The nearest city is 300 km away, sparsely inhabited, and Perth, the capital city of Western Australia is 550 km distant.

The following discussion is grounded in my participation in a range of convivial activities which themselves suggest something of the social aspects of the organisation and running of wildflower shows. Similarly, the following list provides a glimpse of the hive of activity that happens around the seemingly static display of wildflowers. My involvement included spending a day as a volunteer tasked with door-keeping duties, selling merchandise, washing specimen bottles and arranging specimens. I assisted a local picker with the collection of specimens and attended, as part of a large audience, the book launch of *Wildflower Country*, a 'collection of poetry and photography celebrating the 25th Anniversary of the Ravensthorpe Wildflower Show' (Southern Scribes, 2007: back cover). I interviewed two long-term high-profile volunteers, and engaged in informal conversations with several local women also with long-term close involvement in the show. The analysis also draws on committee minutes and publications produced by Ravensthorpe Wildflower Show Inc. (pamphlets, website, book, video and DVD), together with reportage of the wildflower shows in the local newspaper from 2007 to 2009.[1] This study thus privileges 'insider' perspectives and experiences drawing forth a range of local roles and practices around an event which traditionally attracts many more tourists than locals.

Integrated Tourism

Ravensthorpe Wildflower Show is highly successful in attracting tourists. Committee records show attendances ranging from 1000 in the early years to 2000 in more recent years, averaging in the vicinity of 1200–1500 in latter years. A substantial number of international visitors are recorded each year, hailing in particular from New Zealand, the United Kingdom, the United States and Europe, along with consistently a large number of interstate visitors. The majority of visitors arrive via one of 16 Western Australian and interstate coach companies which include a stop in Ravensthorpe as part of their September (if not 'wildflower') tours (Wildflower Show Committee correspondence). As a long-serving committee member noted, 'There is actually quite a circuit of wildflower shows' the majority of which take place in September and October. Indeed, in 2009 the Ravensthorpe Show was one of nine rural wildflower displays (described variously as 'shows' and 'festivals') listed by the Wildflower Society of Western Australia (Wildflower Society of Western Australia, 2009). As another committee member explained, 'The circuit is identified by tourist agencies and coach companies promoting wildflower tours – which are big business for many coach companies'. However, the coach companies 'do not seem to focus on individual wildflower shows'. In fact, the local committee believes that 'despite forwarding information to coach companies it is rarely passed on to the drivers and tour guides'. Instead, in the experience of the Ravensthorpe organisers, coach itineraries are largely governed by pre-booked overnight and lunch stopping places. There is also evidence of what might be called 'private' circuits; one volunteer noted, for example, that there have been groups of keen botanists who visit the Ravensthorpe Wildflower Show regularly from distant Europe and whose itineraries often include other wildflower shows. While 'many visitors from some of the coach tours will omit or cut short their lunch break' in Ravensthorpe in order to visit the Wildflower Show (which is just a few minutes walk from the lunch venue), organisers have observed that 'the people most interested in the flora tend to be those travelling independently'.

Wildflower shows have a long history in Western Australia: in 2009, the Busselton Wildflower Exhibition was in its 84th year (Wildflower WA, 2009), the Mullewa Wildflower Show was in its 22nd year (Mullewa Wildflower Show, 2009) and the Chittering Wildflower Festival was in its

13th year (Chittering Wildflower Festival, 2009), the latter suggesting that wildflower displays in Western Australia are not exclusively a product of the 1980s or earlier. The other Western Australian shows tend to follow similar practice to the Ravensthorpe event with extensive displays of local flora, naming of the species and, more recently, concurrent satellite events. The particular events associated with each show confirm local difference. For example, the Mullewa Show in 2006 and 2009 featured kangaroo stew made by local Indigenous women, whereas the 2009 Busselton show featured a wood turner working with local timbers. The Ravensthorpe Wildflower Show, in addition to its particular associated events and unique species, is prominent among the various local shows for the size of the collection, described locally as 'one of the largest in the world' (Taylor, 2009). The wildflower display traditionally includes a 'Wildflower Show Shop' selling mostly but not exclusively local craft and produce with a wildflower theme. In Ravensthorpe the events vary from year to year, and in this manner, the show is to some extent reinvented (and revitalised) each year. A 'Market Day', introduced for the first time in 2007, offered 16 stalls and displays, free talks from 'environmentalists' on 'nature-based subjects', along with Devonshire Teas. A well-known botanical artist worked on-site and, as noted above, a volume of poetry was launched in honour of this silver anniversary of the show.

Before the Wildflower Show Committee was formed, the Ravensthorpe Wildflower Show was organised jointly by the Ravensthorpe Historical Society and the local senior citizens group. The show continues in the service of these two community groups from which a substantial number of volunteers are still drawn. Participation in larger tourism flows achieves significant financial benefit for these two community groups. The 2009 Visitors' Book lists 1300 attendees and $17,500 was raised from entry fees and the sale of merchandise. In the words of a key organiser, the show is a 'major fundraiser for the senior citizens and historical society. In fact both of those organizations rely fairly heavily on this funding'. The bulk of the merchandise on sale at the show has local origins; these sales in addition support local artists. As this suggests, the annual Ravensthorpe Wildflower Show is an exemplar of integrated tourism (Oliver & Jenkins, 2003: 295) demonstrating clear connections with local resources, products and inhabitants. Its success and longevity derives however from participation in, and also creation of, a range of connections and flows within and between local communities. The Ravensthorpe Wildflower Show, so far, has been able to draw on significant local participation (as opposed to local attendance). As described by one local resident, and confirmed by committee records: 'there are about 50 people who assist in one way or another' with contributions including setting up, staffing door entries, maintaining the display, and cleaning up and packing away.

The show each year thus depends on the considerable, sustained labour of a large number of volunteers. The majority of these volunteers have contributed for many years – some of them since the very first show. Not surprisingly then, 'Most members are aged from late middle-age to getting-on-a-bit!! New, younger people would be very welcome' (H.T., 2007: 19). Not all volunteers are locals; enthusiasts travel from other communities, while yet others, for example, are volunteers from the Western Australian Herbarium who of their own accord regularly assist in Ravensthorpe. In this way, the show is constitutive of wider community networks, traditions and friendships.

Changing Places: Flowers on the Move

Much of the above community volunteer labour is centred on the movement of flowers. The Ravensthorpe Wildflower Show depends for its existence and success on the natural production of flowers (in the wild) and their subsequent movement, as specimens, from 'the field' to the display area. Whereas collecting specimens for the local herbarium requires collectors to follow a precise and carefully documented process, picking flowers for display at the show, in the words of a volunteer, is more a case of 'you just go out and look for flowering plants really. You need enough to display; that's all'. As pointed out, 'none of us started off with other than the most basic of knowledge about plants, so the main requirement is enthusiasm' (Ravensthorpe Regional Herbarium, 2009). This enthusiasm is a key aspect, part of which, as signalled in Lamperd's poem, lies in discovery. Likewise, the poem offers the relationship between 'eager pickers' and 'amateur botanists' as central to the (local) experience of wildflower shows, suggesting a desire to participate in this construction and sharing of local place knowledge, and a hierarchy of knowledge. The extensive yearly display both requires and produces intimate knowledge of what plants are found where and, more specifically, where the best specimens in a given year might be found. The picker I travelled with had a deep knowledge of this 'what' and 'where' for that specific season.

Importantly, the wildflower show provides motivation for and a legitimated mode of being in the local countryside. The collecting, for the many regular pickers, is a 'return' to particular areas, building a diachronic relationship with specific sites supported by rich stores of memories around the presence or absence of wildflowers. The picking is 'authorised' in the sense of conforming to (and thereby overcoming) restrictions to access: 'Unlimited gathering is illegal and collectors obtain licenses to pick' from the Department of Environment and Conservation (Taylor, 2009: 1) and from the Shire. Picking is more than a recreational undertaking: an article in the local newspaper light-heartedly suggests both adventure and conquest:

> Armed with secateurs, buckets and gloves, a battalion of wildflower pickers has been out at dawn for the past week infiltrating the bush armoury of prickles, running the gauntlet of sleepy dugites and fighting their way through blockades of spider webs. (R.G., 2008: 1)

An attendant civic dimension of this work and its direct benefits for the local population is made clear:

> This sacrifice is all for you, the people of the Ravensthorpe Shire, so that you may see in comfort, enjoy en masse, the botanical wonders of our environment. (R.G., 2008: 1)

Not surprisingly, attempts to recruit new volunteers emphasise that 'It's a great way to help the community and learn about our fabulous flowers' (H.T., 2007: 19).

Embedded in the trope of a 'battalion' of pickers, and the attendant notion that the wildflower specimens are won through an organised battle with nature, is a sense of collaboration and sociality. Whether pickers work individually or in groups, there is a clear social element as occurs, for example, in the arrangement of a roster of pickers to cover the region and in opportunities to meet up either at the Senior Citizens Centre or in the field. Withers and Finnegan (2003: 334, 337) in their discussion of 'the role of fieldwork in the activities of natural history societies in Victorian Scotland' argue that it was a significant part of 'the making of local natural knowledge'. Fieldwork, they demonstrate, is also about how knowledge travels; it is about the civic display of a 'locally encountered nature'. Picking for the wildflower show, though also undertaken for pleasure, is principally about the systemic 'field work' collection of 'evidence' to be identified, ordered and publicly exhibited. It is an important part of the making and enactment of a socially valued form of local knowledge of nature. As Withers and Finnegan (2003: 335) find in their analysis, picking for the wildflower show (not least as social event) functions as 'a means of making the social world scientific and the scientific world social'.

An overarching goal of the wildflower show, in the words of a long-serving organizer is to promote Ravensthorpe in 'an eco-tourism fashion'. In 2007, for example, the show included a display by the Department of Environment and Conservation (H.T., 2007). Broadly, the show attempts to 'be an educational tool and to bring to the public something that they don't necessarily get a chance to see in other ways'. For instance, the wildflower show 'helps to illustrate to people the variety of flowers there are in the different families because we display the flowers in family groups so that you have say all the banksias together'. Reinforcing the rural as no longer 'synonymous with agriculture' (Oliver & Jenkins, 2003: 295), the sharing of intimate place knowledge with those who pass through is thus an important aspect of the show, just as it positions Ravensthorpe in a 'new' or 'alternative' role for rural regions in Australia, as steward of natural resources (Stayner, 2005).

At the same time, the display encourages other knowledge flows. For example, in 2007, the Director of the Western Australian Herbarium is locally reported as having 'found two new species on the [Ravensthorpe] shelves. (We'd known about them all the time!!)' (H.T., 2007). While it was not the case that these species were 'unknown to science', as one organiser pointed out, it was in 'the course of a conversation' with the director – a conversation between parties brought together by the show – that the 'anomalous situation where a couple of specimens [in the local collection] appeared different but had been identified as the same by the Western Australian Herbarium' was drawn to his attention. Subsequent taxonomic work confirmed the difference identified locally. In 2008, an ABC reporter examining the food values of native plants attended the show, a professional photographer took studio portraits of the flowers and a Japanese doctor came to study medicinal properties of the plants (R.G., 2008). This external 'expert' attention plays a role in the place-making of Ravensthorpe as 'rich' and 'valuable' landscape. Knowledge of local species is also deployed on behalf of the community; volunteers associated with the wildflower show and the herbarium draw on these resources to provide advice on such things as local reseeding projects (see also Chapter 6). More generally, as an interviewee explained:

> we provide knowledge about where things will grow for the different areas for rehabilitation, for any source, and plants that would be salt tolerant and things of that nature, and also where flowering plants when they occur for beekeepers and things like that.

The well-stocked display demonstrates this local knowledge, providing quantitative and qualitative evidence of a landscape celebrated as 'one of the richest areas of native plants in the world' in which the flowers are 'special because they are unique and diverse' (Craig, 1995: vii). The impressive variety of flowers amassed in the exhibition counters the lack of sweeping displays in the landscape. Similarly, it enacts a temporal compression, a concentration across both place and time as suggested by the following comment offered by a local resident in conversation about the Show:

> For the tourist people who are spending a very short time, it's nice that it's there and it's in one place and that people can look at a whole bunch of stuff that's all local.

This emphasis on the needs of the tourist and the value of the show for this audience is indicative of a wider local perception of the show. In the words of a Shire resident with limited involvement:

> From my perspective, its value is to tourism because from what I've seen, very few of the locals actually go to it because we live here, you know, we see all these flowers, we can go out there and wander through the hills every day of the week if we want to during the wildflower season.

This is not to say that the wildflower show is not intended for or valued by the wider local population; rather, the lack of local attendance is a direct corollary of successfully (re)producing what are seen to be widely available local experiences and knowledge. Through articles in the local newspaper, for example, the committee nevertheless strives to encourage local residents to attend and takes heart from each success as exemplified in the anecdote of the local couple who after living in the area for 20 years finally attended and 'loved it'. Even so, the actual audience for the show, as volunteers concede, is 'mostly interstate and intrastate visitors'.

Committee members note, however, that recent additions to the show, such as the Market Day, are 'primarily a local attraction'. In this way, local support may be more extensively demonstrated. As a further example of this tangential interest, other local groups 'are now tending to link some of their activities to the wildflower show weekend' so that art exhibitions and guided wildflower tours of the region's national parks provide opportunities for local (and visitor) participation in a more extensively conceived wildflower program. Each year the show is reported in the local press maintaining community awareness. Thus, though the number of local attendees may be small, the show focuses community attention on not just the presence of the flowers but also on the extent and value of local knowledge about them and their importance in and to the local community, the broader region and beyond. The annual show articulates a local celebration of the landscape independent of local attendance.

What Name is this?

The activities described above draw on and promote a 'scientific' way of understanding. As a long-standing volunteer commented, the wildflower show has 'developed more scientifically in that first of all it was just a display of wildflowers and then we started putting it in families and definitely trying on the educational angle as well'. A critical aspect has been the scientific classification of flowers and plants:

> a small band of faithful botanists **name all the specimens** that are put on show (all 700+ of them). It's a painstaking business. (R.G., 2008: 1, bold in original)

This 'painstaking' process takes place in the herbarium adjoining the main display area (Figure 10.1), where volunteers spend many hours 'totally absorbed' in 'dissecting flower components and squinting into their individual microscopes' (Taylor, 2009: 2). Recording the number of species collected each year and accurately identifying the correct scientific name and the family of each specimen appears to be a consistent feature of local wildflower shows in general.

The Ravensthorpe Regional Herbarium is central to the acquisition and ongoing availability of skills and resources to make this naming possible.

Figure 10.1 Show volunteers identifying wildflowers (Photo: Robyn Mayes)

The presence of the herbarium was the result of an invitation extended in 1997 to the Ravensthorpe Wildflower Show Committee to participate in the 'Regional Herbaria Programme' (Ravensthorpe Regional Herbarium). Run entirely by volunteers, the herbarium inspired the committee to:

> the lofty aim of collecting every flowering plant in the Shire of Ravensthorpe. [...] There are currently about 3000 specimens in the herbarium – some of them duplicates. (Ravensthorpe Regional Herbarium, 2009)

In the words of a founding member it is 'quite a reasonable collection I think considering the length of time we've been working and we have been lucky enough to have been given grants for computers and a microscope and books and things like that as well'. Although the wildflower show predates the herbarium, and is what precipitated the offer to be part of the regional scheme, the herbarium has become central to the show. Significantly, the herbarium establishes credibility:

> This results from the fact that the volunteers collect two specimens along with the necessary data, and when pressed, one specimen is forwarded to the WA Herbarium. There it is identified and data-based with the result being sent back to us, giving our specimens the necessary provenance. (Ravensthorpe Regional Herbarium, 2009)

This credibility underpins the worth of the work undertaken each year in attempting to correctly identify each plant, just as it underpins the validity of local knowledge in the context of what might be termed the 'scientific' or 'botanical' community.

Engaging Emotions

Alongside this 'scientific' perspective, the show celebrates subjective emotional and aesthetic pleasures and ways of knowing. Lamperd's evocative poem, for example, highlights not only a sense of excitement around the identification of the flowers, but also the social interaction and friendship sustaining the process, and, in its comfortable reference to 'next year', suggests a desire for and sense of enduring tradition. Accounts of the show foreground the smells and ambience of the flowers through descriptions which emphasise, for instance, the 'heady perfume' and the 'blaze of colour' (Taylor, 2009: 2). The ordered displays are each year accompanied by massed displays of the flowers (see Figure 10.2).

A unique feature of the 25th anniversary show, *Wildflower Country*, as a whole honours a quarter-century of local emotive experiences of and relationships to a landscape presented as intensely local. The inclusion of Lamperd's poem, written some time in the first few years of the show

Figure 10.2 Wildflower display, Ravensthorpe Wildflower Show, 2007 (Photo: Robyn Mayes)

(pers. comm.), signifies an ongoing emotional tenor of each show. Dedicated to 'Jim McCulloch OBE', who was responsible, in the face of considerable local resistance, for the first show in 1982 (Craig, 1995), the volume is the work of Southern Scribes, a long-established creative writers group whose members live 'in or near the towns of Ravensthorpe and Hopetoun' (Southern Scribes, 2009). Each poem in the book is accompanied by a photograph; images of local flora predominate and are often the subject of the poems they accompany, with a handful of landscape photographs including one of local residents enjoying a picnic. The photographers are acknowledged in an index which provides a title for each photograph. These titles include the common name of the plant in question along with its botanical name: for example, 'Qualup Bell – Pimelea physodes'. Vernacular knowledge, subjective experience and scientific naming are thus brought together in highly place-specific ways. The description on the back of *Wildflower Country* makes the connection to the local explicit: 'you don't need to have visited the region to enjoy this book, but it will make you want to experience the colours and textures, the tastes and perfumes that belong only here' (de Garis, 2007: back cover).

The book celebrates experiences of the local countryside as shaped by age and gender ('Grandmothers Have Time'), reflects on place change as a sense of loss and on the importance and origins of stewardship

('At Starvation Bay') and grapples with the ways in which progress/civilisation impinges on the natural landscape ('Dampiera sacculata' and 'Daybreak Highway 40'). Emotions such as pride, pleasure, joy, friendship and belonging infuse the senses of place under the rubric of 'Wildflower Country'. The book articulates what Jones (2005: 205) has argued is an 'inevitable' conjoining of 'emotion, memory, self and landscape', facilitated and brought to the fore by the (repeat) occasion of the wildflower show. Emotions are fundamental to making meaning both in terms of 'producing a *meaningful* world, a world worth caring about' and shaping 'our modes of "being-in-the-world"' (Smith, 2005: 220, 291). These emotions have not only spatial but also temporal referents, and in coalescing 'around and within certain places' point to 'relational flows, fluxes or currents, in-between people and places rather than "things" or "objects" to be studied or measured' (Davidson *et al.*, 2005: 3). The annual wildflower shows are thus about more than the taxonomic ordering of flowers and visual objectification, though these are to some extent privileged. For example, the mixed, massed floral displays, though an important part of each show, are made from plants remaining *after* the specimens for display have been selected.

The Centrality of Flowers

The flowers – as specimens and aesthetic and sensory objects, and/or emotional loci for memories – are constituted by and in turn shape the show in its multiple concurrent and interlinked configurations as a natural, social, cultural and historical event. On one level understandable as a 'natural' state of affairs, the centrality of the flowers is something the local committee consciously strives for. Though 2007 was the 'silver anniversary' and plans were in hand to celebrate the show's origin and history, the flowers were kept firmly at the heart of the show; according to a local organiser speaking during the planning for the 2007 event, 'it probably won't be a very big [history] display given that we still want to run the wildflower show as a wildflower show'. Likewise, though the shows are effective fundraisers, fundraising is not offered by those involved as a motivation. Rather, fundraising is an attendant, subordinate, benefit. The focus for local participants is also the flowers. To return to the opening poem: 'They arrive', and the show can begin. The flowers are important also on a deeply personal level: one regular volunteer observed: 'I do it for the flowers'. As also expressed in the poem, names may change but the flowers will still be there next year. The event itself, as discussed above, is in the service of wildflowers, not least in terms of educating people of their importance and value.

The centrality of the flowers is highly physical: the flowers colonise the display space. The Senior Citizens Centre is transformed into a display/festival place 'filled' by the arrival of the plants. As one description has it, the 'treasured cargo was carried in, filling every corner with colour, pollen and the wonderful smell of the bush' (R.G., 2008: 2). Each year, the Senior Citizens' Centre is appropriated as a site in which unruly nature – the trays of unsorted flowers are confined to the kitchen area (see Figure 10.3) – is transformed into an orderly arrangement (see Figure 10.4). Display stands and hundreds of bottles are brought out to best exhibit the specimens. Banners, signs and laminated specimen name cards used in previous years are unpacked and reused. The yearly display of wildflowers is thus a labour of translation and of distillation in which local place knowledge and its corporeal experience travels from 'the countryside' to a place in turn transformed for and by their arrival. Visitors are invited to examine, learn about, photograph and in general marvel at the flowers in a controlled fashion in terms not only of the grouping of species but also the flow of movement and activity encouraged by the rows of displays, the provision of magnifying glasses for closer inspection and the playing of a documentary on the wildflowers of the Ravensthorpe area produced by the show committee.

In both the Senior Citizens Centre and in the countryside, the flowers are powerful non-human actants in the dynamic production of these places.

Figure 10.3 Unsorted wildflowers, Ravensthorpe Senior Citizens' Centre (Photo: Robyn Mayes)

Figure 10.4 Specimens on display, Ravensthorpe Wildflower Show (Photo: Robyn Mayes)

To paraphrase Cloke and Jones (2001: 655), the local plants bring the unique creativity of being able to produce flowers in the first place. The flowers are not always compliant; the number and quality of specimens vary from year to year thus informing the specificities of each year's exhibition. As noted above, those involved over many years in the work of picking and of display engage in 'repeated encounters with places and complex associations with them' which 'serve to build up memory and affection for those places, thereby rendering the places themselves deepened by time and qualified by memory' (David Harvey cited in Cloke & Jones, 2001: 651). The Ravensthorpe Wildflower Show in its interrelated and nuanced articulations examined here enacts what Cloke and Jones (2001: 655) in their analysis of West Bradley Orchard describe as 'a deep hybridity of people, nature, and technology which is embedded in a complex of networks, but which also has a time-thickened, place-forming dimension'. In this instance, wildflower shows are the central interconnecting and seemingly neutral category, mechanism and object motivating and enabling this local hybridity.

Assembling Place

Though the Ravensthorpe Wildflower Show is the product of a historically entrenched, highly specialised interest in the biology of flowers, its yearly conduct is deeply embedded in the local community and landscape. A complex undertaking, with clear material benefits for two community groups, the show points to the substantial role integrated rural festivals play in the generation and revitalisation of diverse local senses, experiences and knowledge of a place. Importantly, this case study foregrounds the way in which the place-specific knowledge reproduced by this particular type of show interweaves scientific/objective and emotional/subjective knowledge. The show focuses attention on wildflowers offered not only as objects of scientific knowledge, but also as central to deeply personal experiences and memories of the place.

The production of place functions on several scales including within broader networks of 'place' as encoded in the (re)invention each year of Ravensthorpe as part of a 'circuit' of wildflower places, characterised as much by similarity with other shows as by difference. The annual show offers Ravensthorpe not only as a place of wildflowers, as a wildflower country, but also as a place to see and know wildflowers. Concomitantly, the show orders and intensifies the experience of wildflowers. Though wildflowers are central to the show, which depends upon their yearly production in 'the wild', amassing specimens according to family classifications imposes an order antithetical to their existence in 'the wild'. The result of extensive field work, itself mobilising and confirming extensive local place knowledge, the specimens on display make highly visible a seasonal transformation less noticeable in the landscape. Offered as typical examples of flowers in the wild, the specimens function as a powerful locus of 'nature/culture'.

Understanding the Ravensthorpe Wildflower Show as 'assemblage', and therein recognising that humans, non-humans and texts can 'potentially change the course of events' (Hinchliffe, 2005: 195), foregrounds the complexity of this (and other rural) festival practices; the Ravensthorpe Wildflower Show productively assembles flowers, community, senses and experiences of place, poems and other written and visual texts and tourists in a manner in which each are also actants. This particular rural festival, and perhaps wildflower shows more broadly, suggests an ongoing rural negotiation of a place as formed in and through human and non-human interaction.

Acknowledgements

My heartfelt thanks to Ann Williams, Helen Taylor, Merle Bennett, Enid Tink, Gillian Smith and Barbara Miller-Hornsey for sharing their considerable knowledge with me, for making me feel so welcome, and for reading drafts of this chapter. I take full responsibility for all errors of fact. A warm thank you to Laurel Lamperd for permission to reproduce her poem 'Wildflower Show'.

Note

1. Much of this research was undertaken as part of a two-year independent ethnography of the Shire of Ravensthorpe funded by the Alcoa Foundation's Conservation and Sustainability Fellowship Program hosted by Curtin University.

References

Australian Bureau of Statistics (ABS) (2007) 2006 census tables: Ravensthorpe (S) LGA. On WWW at http://www.omi.wa.gov.au/WAPeople2006/Wa/Regional/Ravensthorpe.pdf. Accessed 19.06.10.

Chittering Wildflower Festival (2009) Chittering Wildflower Festival 2009 – Online document: On WWW at http://www.chittering.iinet.net.au/promo%20wildflower%20and%20wine%20festival.htm. Accessed 19.06.10.

Cloke, P. and Jones, O. (2001) Dwelling, place, and landscape: An orchard in Somerset. *Environment and Planning A* 33, 649–666.

Conradson, D. (2005) Freedom, space and perspective: Moving encounters with other ecologies. In J. Davidson, L. Bondi and M. Smith (eds) *Emotional Geographies* (pp. 103–116). Aldershot: Ashgate.

Craig, G.F. (1995) *Native Plants of the Ravensthorpe Region*. Ravensthorpe: Ravensthorpe Wildflower Show Inc.

Davidson, J., Bondi, L. and Smith, M. (2005) Introduction: geography's 'emotional turn'. In J. Davidson, L. Bondi and M. Smith (eds) *Emotional Geographies* (pp. 1–16). Hampshire: Ashgate.

de Garis, J. (2007) Back cover. In Southern Scribes (eds) *Wildflower Country*. Ravensthorpe, Western Australia: Southern Scribes.

Hinchliffe, S. (2005) Nature/culture. In D. Atkinson, P. Jackson, D. Sibley and N. Washbourne (eds) *Cultural Geography: A Critical Dictionary of Key Concepts* (pp. 194–199). London: I.B. Tauris.

H.T. (2007) Over for another year. *Community Spirit* 19 (18), 19.

Jones, O. (2005) An ecology of emotion, memory, self and landscape. In J. Davidson, L. Bondi and M. Smith (eds) *Emotional Geographies* (pp. 205–218). Hampshire: Ashgate.

Mayes, R. (2010a) Doing cultural work: Local postcard production and place identity in a rural shire. *Journal of Rural Studies* 26, 1–11.

Mayes, R. (2010b) Postcards from somewhere: 'marginal' cultural production, creativity and community. *Australian Geographer* 41, 11–23.

Mullewa Wildflower Show (2009) Mullewa Wildflower Show. On WWW at http://www.about-australia.com/events/western-australia/australias-coral-coast/events/exhibition-show/mullewa-wildflower-show/. Accessed 19.06.10.

Oliver, T. and Jenkins, T. (2003) Sustaining rural landscapes: The role of integrated tourism. *Landscape Research* 28, 293–307.

Ravensthorpe Regional Herbarium (2009) Ravensthorpe Regional Herbarium. On WWW at http://www.wildflowersravensthorpe.org.au/herbarium.html. Accessed 19.06.10.

R.G. (2008) Ravensthorpe Wildflower Show – Don't miss it! *Community Spirit* 10, 1.

Smith, M. (2005) On 'being' moved by nature: Geography, emotion and environmental ethics. In J. Davidson, L. Bondi and M. Smith (eds) *Emotional Geographies* (pp. 219–230). Aldershot: Ashgate.

Southern Scribes (2007) *Wildflower Country*. Ravensthorpe, Western Australia: Southern Scribes.

Southern Scribes (2009) Welcome to Southern Scribes' Homepage. On WWW at http://www.southernscribes.com/index2.html. Accessed 19.06.10.

Stayner, R. (2005) The changing economies of rural communities. In C. Cocklin and J. Dibden (eds) *Sustainability and Change in Rural Australia* (pp. 121–138). Sydney: University of NSW Press.

Taylor, H. (2009) Wildflower Show gains international recognition. *Community Spirit* 19, 1.

Wildflower Society of Western Australia (Inc.) (2009) About our society. On WWW at http://members.ozemail.com.au/~wildflowers/Intro.htm. Accessed 19.06.10.

Wildflowers WA (2009) Wildflower events. On WWW at http://www.wildflowerswa.com/en/Wildflower+Events/default.htm. Accessed 19.06.10.

Withers, C.W.J. and Finnegan, D.A. (2003) Natural history societies, fieldwork and local knowledge in nineteenth-century Scotland: Towards a historical geography of civic science. *Cultural Geographies* 10, 334–353.

Part 4
Reinventing Rurality

Chapter 11
Elvis in the Country: Transforming Place in Rural Australia

J. CONNELL and C. GIBSON

Like many other small towns in inland Australia, Parkes (in New South Wales) has gone through tough times in the past two decades. Restructuring of agriculture and drought brought economic challenges, the loss of activities such as banks and shops and a slowly falling population. In this century, the pattern of decline in Parkes has been arrested and the establishment of an exceptionally successful festival has played a part in that. Indeed the Parkes Elvis Revival Festival demonstrates how a small, relatively remote place can stage a festival that generates substantial economic benefits, fosters a sense of community, seemingly against the odds, and in doing so has gained nationwide notoriety and publicity without any particular local claim to musical heritage. The festival represents about as narrow a rationale for an event as can be imagined – the legendary performer is long dead, and festival visitors arrive to see mere impersonations of the original. Yet the festival has invigorated the town, attracted loyal, repeat visitors and brought a community together on an otherwise hot and dusty weekend in the tourist off-season, because it is well-organised, slightly weird, in a friendly town and, above all, fun.

This chapter discusses how an unlikely festival overcame adversity and local opposition to become one of the most famous festivals in the country. Not only has it gained national prominence but, along with 14 other Australian festivals, mostly metropolitan, it is listed in Frommer's *300 Unmissable Festivals Around the World* (2009) where it is also distinguished as setting (in 2007) a new record for the most Elvises (though the plural form is usually referred to in Parkes as Elvi) in one place.

The chapter is based on repeated visits between 2002 and 2010 to undertake research in collaboration with the organisers of the Elvis Revival Festival, involving surveys of businesses, residents and visitors and interviews with local tourism promoters, local government representatives, tourists and families hosting home stay visitors. Across nearly a decade we have tracked how a small place with few economic prospects has created a tourism resource, and subsequently captured national publicity, through a festival based around the commemoration of the birthday of Elvis Presley, a performer who had never visited Australia, and certainly not Parkes, and had no links to the town. Indeed, Elvis rarely left America. The Parkes Elvis Revival Festival demonstrates how 'tradition' can be constructed in rural places (rather than being innate), how small places can develop economic activities through festivals, and create new identities, though constantly contested.

Elvis Comes to Parkes

Parkes is a small New South Wales country town of about 9600 people 350 km west of Sydney. Like many other inland country towns, it had lost population (4% between 1996 and 2006), had higher than average unemployment rates and low levels of participation in the labour force (43% of the total population), with a population increasingly dominated by those of retirement age. It has long been a service centre in Australia's wheat-sheep belt, though the North Parkes copper mine provides economic diversity, and it is a significant rail centre. However, other than its historic radio telescope ('The Dish'), a vital link in the 1969 Apollo moon landing (which became, in 2000, the subject of a popular Australian feature film of the same name), Parkes has little in the way of visitor attractions.

The establishment of the Elvis Presley Festival in Parkes was entirely the result of a chance local whim, when a couple of local people devoted to the memory and music of Elvis, proposed the idea to council members, as recalled by committee member, Neville Lennox, who later formally changed his name to Elvis Lennox:

> It was Bob and Anne Steel up at Gracelands restaurant. They're big Elvis fans and they own the restaurant. They were just having a bit of a talk to the right people at the right time, at one of their functions. They were councillors and they said, 'Well there's nothing going on, nothing celebrated that time of year. Elvis's birthday's the eighth. Come along to the next council meeting, we'll put it to the board'. It just evolved from there. (Interview, 2004)

Parkes happened to have a club and restaurant called Gracelands, and a small group of committed Elvis fans willing to organise an event. This suited the pragmatic aim of the local council of the time, namely to improve summer tourism, though there was no great excitement about hosting the event. An Elvis Revival committee was subsequently formed and, in 1992, what was a very small group of local fans decided to stage Australia's first Elvis festival. The first Elvis Revival Festival was held in January 1993, coinciding with Elvis' birthday. It attracted about 500 people from as far as Adelaide, Melbourne and Sydney, and set the theme for those that followed, with Elvis and Priscilla (Elvis Presley's wife) look-alike competitions, a street parade with vintage cars, shop window displays of memorabilia, Elvis movies at the cinema (since closed), and concerts, one of which was at the Gracelands Club. Indeed the fortuitous presence of a Gracelands Club had been one factor convincing organisers that Parkes was the appropriate place for the festival though eventually the restaurant no longer had the capacity to hold the crowd, and entertainment and people spread to other pubs and clubs.

The first festivals were largely ignored by the local media as inappropriate for a country town or trivial (despite the dearth of news in midsummer), and that exclusion has only partly diminished. By contrast, the national media have regularly covered the Festival, invariably because of its curiosity value, but also as a result of what were seen as ludicrous claims by the organising committee that it wanted Parkes to become the 'Elvis capital' of Australia. For Bob Steel, then chair of the organising committee, the lead-up to the first festival hinted at such national publicity:

> We have been overwhelmed with the attention this festival is receiving. For example, even the *Melbourne Truth* ran an article on the festival, suggesting that we could become the Elvis capital of Australia. Newspapers, television and radio stations have all been giving the festival plenty of coverage and if nothing else, it has certainly given Parkes publicity. (Quoted in the *Parkes Champion Post*, 8 January 1993: 5)

Ironically, this kind of national coverage, and its celebration of tackiness and kitsch, has probably drawn most visitors, as typified by one picture in the *Sydney Morning Herald* (Figure 11.1). Indeed the coverage of the Festival in every year has focused almost exclusively on the multiple, gaudy jump-suited Elvis.

The festival once began on the Friday night of the weekend closest to Elvis' birthday (8 January 1935); although since 2008 it has been drawn out to earlier in the week and longer into the next week. It has usually involved dinner and various forms of Elvis entertainment at Gracelands (although that club recently closed), with all participants encouraged to dress in appropriate annual themes: cowboy, speedway, Hawaiiana; usually linked to Elvis movies. Saturday sees the street parade of vintage cars and motorbikes (and vintage Elvis impersonators), with market stalls (ranging from memorabilia – rarely 'real'– to country handicrafts) in the main park area. The park is the venue for the main sound and look-alike

Figure 11.1 Parkes, according to the *Sydney Morning Herald* (January 2003: 8) (*Source*: Brennan-Horley *et al.*, 2007: 75)

competitions – Elvis, Priscilla, Lisa-Marie (Elvis' daughter) and Junior Elvis – and the day concludes with several feature performances in different local clubs by touring professional Elvis impersonators. The highlights of the Sunday are the highly attended Gospel Church Service, further competitions and performances and the unveiling of a new plaque on the Elvis Wall (at the park where the Festival first began) to commemorate another 'legend' of Australian rock 'n' roll music (often one of the previous night's top-billing performers). The wall itself surrounds gates that are a replica of the gates of Presley's Graceland mansion in Memphis. A talent contest with more diverse themes brings the festival to an end as most visitors return on Sunday night often over considerable distances. A special train (the Elvis Express) runs from Sydney (Figure 11.2), with the support of CountryLink rail which has become the main sponsor of the festival. Many of those who use the train are dressed as Elvis or Priscilla and CountryLink provides its own Elvis impersonator to perform in the carriages. By 2010, the Elvis Express had eight carriages and almost 400 passengers: the physical capacity of the line. A second train from Melbourne has been planned.

On some occasions Elvis movies have been shown and the local lawn bowling club has urged visitors to 'kick off your blue suede shoes' and have a game. An Elvis celebrant is available for couples to marry or renew

Figure 11.2 The CountryLink Elvis Express (*Source*: State Rail, promotional material 2005)

marriage vows during the weekend, and that has become extremely popular. Elvis buskers occupy all the street corners (and there is a prize for the best), and the Private Collection of memorabilia of Elvis Lennox – with a pink Cadillac parked in the driveway – is open to visitors. The emergence of Parkes as home of the Elvis Festival played a pivotal role in influencing Neville to take on the icon's name:

> I prefer Elvis to Neville, me original first name. After the first two years of competition here in the look-alikes – I won that in 93, 94 – and walking up the street or down the street, whichever the case is and you hear people yell out across the street at ya 'g'day Elvis' and that. And I said, 'ya know, that would be an idea'. So I put it to me mother, asked her permission to do so and she said 'you go ahead and do with it what you want'. And I said, 'thankyou very much'. Paid 75 dollars and had it legally changed. (Interview, 2004)

An avid collector, Lennox amassed a formidable amount of Elvis paraphernalia, some of which comes from a personal trip to Memphis in 1997, the same year as his name change (Brennan-Horley *et al.*, 2007), and he has been a stalwart of the organising committee since the Festival began.

In its second year, the festival brought visitors from further afield, including Western Australia and Queensland, and added a clambake at Gracelands, with sand and surfboards brought in to transform the car park. The Parkes Tourism Promotions Officer heralded it a success, and conceded that it had become an integral part of the annual events calendar.

Although interest grew steadily, the organisation of early festivals was a struggle, and even the elements conspired against success. In two of the first four years, bush fires prevented visitors leaving coastal New South Wales to travel inland and then floods cut off the town. The small number of visitors suggested that the Festival might founder. However the local rugby team decided to support the Festival as a fun event and began what has become a tradition of dressing up as tacky Elvis look-alikes, in jump suits from the late Las Vegas years. Since then for the entire period of the Festival the town, and especially the venues, is seemingly awash with Elvis impersonators (Figure 11.3) which for many creates a colour and atmosphere that is the hilarious highlight of the Festival. Local support and this new image first got the Festival through difficult times.

Even so for most of the 1990s the Festival barely survived; leadership was lacking and local event management skills were few. As Kelly Hendry, the Parkes Tourism Manager, explained:

> It started off small and started to grow. The word started to get out and the media coverage got out about the festival but I guess the lack of resources and lack of skills among the committee and just a few different things, and lack of support from the community saw numbers start to dwindle and the festival nearly fell over a couple of years ago. That's when the tourism board got back on board again. (Interview, 2004)

For its first decade the Festival struggled to galvanise support amongst people who perceived it as tacky, inappropriate for a respectable country town, with no local relevance and taking place in the hottest month of the year when temperatures were normally above 30 degrees and 'escaping' to the coast was almost essential. Many people saw the 'Dish' as a more appropriate symbol of the town, and an American performer who had probably died of a drug overdose as at best irrelevant and at worse degrading.

However, external media coverage never flagged and each year new visitors arrived for the Festival. As the Parkes Tourism Board gradually

Figure 11.3 Elvis impersonators busk outside the Royal Hotel, Parkes, 2010 (Photo: John Connell)

warmed to it publicity increased and numbers too gradually increased. As Kelly Hendry observed, 'it's got that uniqueness and no one else is doing it' (Interview, 2004). By the early years of this century the street parade was drawing a crowd of around 2500, with one or two hundred at most of the commercial events, and more than 500 estimated to have come from outside the town. In 2006, organisers estimated that over 5000 people participated in the festival. Since then it has grown and estimates for 2010 suggest that as many as 10,000 had come into Parkes at the peak time on Saturday morning.

For the first time in 2002, media coverage became international, with Japanese film crews setting up noodle tents to feed hungry Elvises. The kitsch element of the festival was growing too. While the Parkes Shire Council eventually provided financial support, and it is now partly locally funded and sponsored, it is run largely voluntarily by a committee of locals, tourism promoters and Elvis fans, with all profits going to local charities. In recent years, further financial support has come from the New South Wales State government and major sponsors. By 2010, there were ten of these, including the regional Rex Airlines, Country Energy and North Parkes Mine, and a host of minor sponsors. In 2004, it was officially supported for the first time by the New South Wales State government, under the Regional Flagship Events Programme, with the Minister for Tourism observing that

> What the Parkes Elvis Revival Festival does for regional New South Wales is act as a flagship by attracting more tourists. The Festival is always the highlight of the New Year in central New South Wales. (Quoted in *Parkes Champion Post*, 17 January 2005: 3)

By the 18th festival in 2010, which coincided with what would have been Elvis's 75th birthday, there were some 140 distinct events spread over five days (ranging from Bingo with Elvis and Hunka Hunka Breakfast with Elvis through dozens of musical events to the Elvis Golf Challenge), approximately 400 or 500 Elvis impersonators (not all of whom, fortunately, sung), and 10,000 visitors, half of whom were from the nearby region. By then the Sunday morning Gospel Service had become the single largest event with more than 2000 people. What had begun in the local Baptist church now took place in the giant Woolworth's car park, the only 'venue' in town that was large enough. Five years earlier, the key parts of the Festival – the main open air stage and markets – had moved from a peripheral location to the very centre of the town, where they were more accessible to shops. Elvis had come to town.

Visitors: Who, Where from and Why Elvis?

Tourist numbers have become considerable, and now larger than the population of Parkes. Festival visitors were first surveyed in 2003 (125 respondents) and have been since every year, providing data on their demography, expenditure patterns, transport arrangements, accommodation type, motivations to visit and their experiences in Parkes.

The age of visitors to the festival has always been somewhat older than that at other music festivals, although intriguingly it is almost identical to that at the very different Opera in The Paddock (an annual event in a field several kilometres outside Inverell, northern NSW), at which we have also conducted research. The Elvis Festival is dominated by people from the 45 to 65-year-old cohort, who made up over 60 percent of all visitors in every year since we started surveying visitors. In 2010, some 84% of respondents were aged over 45. This distribution was unsurprising, reflecting the considerable popularity of Elvis with people who experienced their youth when Elvis was alive and active as a performer. That there was some 'aging' between 2003 and 2010 may have partly reflected return visiting. Younger people were fewer, and those who were present tended to see the event as a fun 'kitsch' or 'retro' event, rather than about nostalgia or reminiscence.

Festival visitors came from a range of occupational backgrounds. In 2003 and 2004, the largest group were professionals, a group well known for their propensity to travel and for their high levels of attendance at festivals, followed by tradespersons, retirees, and managers and administrators. By 2010, the largest group was now retirees, again suggesting return visits, but professionals and trades people were well represented.

In 2003, as many as 80% had not attended an Elvis Revival Festival before, but respondents enthusiastically said that they were likely to return to the festival. Of those who had attended previously, most had visited in several consecutive years – a measure of the presence of 'devotees' at the festival, for whom the Elvis festival was much more than mere entertainment. By 2010, some 53% had not attended before; return visits had become much more important. Of those who had been before a dozen had been more than 10 times. A couple had been to every one.

In early years, word-of-mouth and newspaper advertisements were the most common ways that visitors found out about the festival, with a modest rise in visits to the festival website in 2004, perhaps a reflection of increases in the numbers of younger people attending. By 2010, word of mouth was again most important but over a third of visitors claimed 'prior knowledge': the Festival was now well known. Direct advertisement was less crucial.

Much like other festivals, most visitors came from nearby. Many participants were from Parkes itself and more than half were from regional New South Wales, especially the central-west region that includes Parkes. Of the 30% who were from further afield most came from Sydney and fewer from other states. However, it was these more distant visitors who were more likely to stay several nights, to spend substantial sums of money especially on evening club performances, to be 'serious' Elvis enthusiasts (often members of Elvis fan clubs, and rock 'n' roll clubs, such as Lithgow Workers Rebel Rockers Dance Club) and more likely to come repeatedly. By 2010, the Elvis Express had become a reunion of old friends, singing, dancing and reminiscing, and eating Elvis Cupcakes and Love me Tender Chicken, throughout the journey.

Most people attending the Festival not surprisingly came for fun, relaxation and a sense of community. When prompted on their experiences, well over 90% had enjoyed the entertainment, country hospitality and music. Between 2003 and 2010, that never changed. A little less important as a general rationale was 'Because I'm an Elvis fan', so that the Festival involved many people who were there because it was a fun weekend, for whom a generalised nostalgia was sometimes of significance, but the actual theme was not necessarily the key to participation. There was also a large minority for whom being an Elvis fan constituted the main reason for participation, and who eagerly anticipated the festival year after year.

Many saw the festival as an opportunity to let their hair down: 'I put all my Elvis things on this morning. I can't wear it around Warwick [Queensland] because people would think we were a bit queer, but here we can express ourselves' (Interview, 2007). For others nostalgia dominated: 'It brings back your youth. And it's just the joy you experience now. I play Elvis music every day. Not many days go by that I don't actually sing Elvis music. I spend about five hours a week doing Elvis things' (Interview, 2010). For such people, travelling to Parkes was something of a pilgrimage.

Return visitors were more likely than others to be Elvis fans, enjoy the music and enjoy spending time with friends and family. Such testimonies indicate the manner in which festivals – even the most seemingly esoteric or incidental – transcend daily life and bring a range of meanings to individual lives. For a handful of fans, the visit to Parkes was akin to pilgrimage (cf. King, 1994), albeit a pale reflection of the trip to Graceland in Memphis, but the closest that Australia can offer. Some visitors suggest the presence of 'postmodern' tourists (or post-tourists), visiting Parkes for the humorous and kitsch ('everything was sensational, baby! uhh huh huh!'; 'eating at Gracelands – wow – I've been to Gracelands!') (Figure 11.4). For particularly committed fans of Elvis, there is essentially no other means of expressing such devotion, without lengthy and expensive travel to America.

The Town Becomes Full

On our first visit to the Festival in 2002, it was possible to arrive in Parkes on the Saturday morning in time for the parade and book into a motel for Saturday night. Hotels and motels still had some spare capacity. That quickly ended as numbers grew. Parkes had succeeded so well from the Festival that, by 2006, it effectively reached the limits of local

Figure 11.4 'Eldest Presley', Parkes Elvis Revival Festival, 2010 (Photo: John Connell)

accommodation, and Parkes has 13 motels with about 1000 bed spaces. Not only had Parkes become full, but towns such as Forbes, some 35 km away, were also full. Dubbo, even further away, was almost booked out by the end of the decade. Routinely by the end of one festival signs have gone up outside all the motels that Parkes is already booked for the following year, a further indication of the strength of return visiting.

In 2004, Parkes made the decision to establish a 'tent city' on the edge of the town where visitors could hire tents and have access to basic facilities, and where caravans could also be parked. That was modelled on the experience of the Tamworth Country Music Festival where a tent city had long been successful. The same company that established the Tamworth tent city developed the Parkes tent city, and revenue mainly accrued to the operator.

Two years later, Parkes decided to establish home hosting modelled on similar schemes in the larger NSW towns of Gunnedah and Bathurst, in association with festivals in those places. The intention was to meet continually expanding demand, ensure that more revenue from accommodation remained in Parkes and provide a friendly and homely experience. Other NSW towns, including Moree and Tamworth have subsequently adopted the scheme.

This home stay system involves local residents with spare bedrooms offering their homes as accommodation, and in return they receive 50 dollars per guest per night. The majority of the money made from home hosting goes either to the hosts or to the festival itself, helping the income remain in the town. Hosts provide a continental breakfast and a ride to and from the train station as guests arrive and depart. In 2006, the first year, just four homes and 15 guests stayed. By 2010, there were 125 homes and 547 guests (and 1561 bed nights; most visitors stayed for three nights). In other words, home hosting provided a third of the formal beds in Parkes. The cost to the guest was $66 per bed per night, of which $50 goes to the host, $10 to the Elvis Festival committee and the remaining six dollars used to partially cover the expenses of the home hosting programme coordinator. Both hosts and guests were enthusiastic about the programme and many guests returned to the same host in subsequent years. Almost all the hosts joined the programme to support the local committee and the town; just a couple were in it 'for the money'. While most hosts were genuinely altruistic and enjoyed meeting people the income generated was valuable at a time of economic stress. As one host said: 'it's been the savior of the town with the drought' (Interview, 2010). Revenue went directly into the hands (and pockets) of local residents, increased interest in hosting and widened local support for the Festival.

The Economic Impact of Elvis

Like many small festivals the Elvis Festival made no money in its early years, and that in itself meant that local support was subdued or non-existent. But by the 2000s, that was changing rapidly. Visitor surveys in 2004 indicated that economic impact of the festival had already become considerable. Visitors then spent an average of A$440 per person over the festival weekend, translating to an injection of over A$1.1 million into the local economy. Accommodation (averaging A$142 per person), food and drink (A$134) and entertainment (A$51) were the most common forms of expenditure, with smaller amounts spent on souvenirs (A$43) and other services such as fuel (A$28). For a town of its size, that expenditure was considerable; both because there were relatively few services in some categories while multipliers spread that revenue through the local economy. By 2010, the direct visitor expenditure contribution of the festival to the town was over $3 million. Moreover the impact of the Festival was felt much further away, in towns such as Forbes and Dubbo where people stayed, in campsites, motels and caravan parks in a number of towns nearby or en route, and likewise in petrol stations and cafes far away from Parkes itself. There are significant regional impacts.

Some of that local expenditure went to the local market stalls. In its earliest years, there was virtually no commercial presence, and even in 2002, there were merely a dozen stalls doing a desultory business selling local goods. By 2008, the number of market stalls had passed a hundred and the main park was so crowded that numbers had to be cut back to 70 in 2009 to allow crowd movement. Stalls now sold local rural goods – honey, jams, soaps and handicrafts – while some stallholders came from inter-state as part of a national circuit (Figure 11.5). Only a handful sold Elvis memorabilia. Local businesses – the Rotary Club, the Lions Club, schools, the fire

Figure 11.5 Stallholder, Parkes Elvis Presley Revival Festival 2010 (Photo: John Connell)

brigade and so on – had their own stalls and barbeques that did good business. Indeed, most things sold well. As one visitor observed:

> You've got Elvis wine, Elvis beer, Elvis tooth brushes, there's heaps of stuff – it's really tacky ... the tackier it is the better it is ... I mean people are buying 45 foot Elvis rugs ... which is classic behaviour at a festival ... the details are irrelevant. People consume all this memorabilia because people are in the spirit of it and that's what a festival does, it changes your behaviour. (Interview, 2007)

The majority of formal Parkes businesses recognise that trade increases during the Festival, to the measurable extent that across years we surveyed businesses (2004–2008) a quarter of businesses put on extra staff over the weekend, adding a total of between 30 and 50 jobs to the town. Predictably, restaurants, cafes, clubs and accommodation facilities accounted for the bulk of new shifts created. These businesses were also those with the highest dependency on local suppliers and labour. The festival improved employment multiplier impacts by generating extra work in those activities that, in turn, are most closely embedded in the local economy rather than others that rely on goods and services (such as books and clothes) imported from state capitals and beyond. Over time, the businesses that benefited most from the influx of visitors stayed open much longer; Saturday afternoons and even Sundays were much less 'dead' than in earlier years or on other weekends, and further multiplier effects ensued. Elvis had been taken on board.

Whose Town?

Until quite recently, Parkes rarely mentioned the Festival in any of its standard tourist publications, preferring to advertise itself as the town with 'The Dish', and as a prominent regional commercial centre. Its longstanding tourism brochure simply ignored the Festival. Only since 2007, has it been officially mentioned.

In the 2004 survey, local businesses were questioned about the appropriateness of the festival as a marker of place identity, as opposed to other options such as 'The Dish'. Opinions were divided. The majority (62%) were strongly supportive, most of the remainder were 'mildly supportive' while some 5% expressed no support at all for the festival. Over 80% of businesses agreed that the festival had a positive impact on publicising Parkes as a tourist destination, yet over 65% either mildly or strongly favoured 'The Dish' as a source of more appropriate imagery.

Business reaction to the Festival was mirrored in broader community sentiments. But as the Festival grew these too became more positive. Interviews with some thirty local residents immediately after the 2009 Festival found either substantial support, or mere indifference, with little significant hostility. For some it was 'the best thing that has ever happened here'; an 81-year-old noted 'it's great for the town and the people'. Another woman argued: 'Those who only think of the Dish live in the past and we have to be more creative now'. While some preferred to stay away from the crowds or Elvis was 'not my thing', just one 61-year-old woman was bitterly opposed: 'I hate it; it closes off the main street, there's no access to businesses. Since the first one, really rude people came from Sydney. Prices go up. We should promote the dish and our good restaurants, but we're under the thumb of the council'. By contrast, perhaps most convincing of all was the woman who explained

> I hated it when it first started. It was ridiculous and stupid and wasn't the image that was at all appropriate to our town. But over the years I watched and could see that it was making money and wasn't so bad. Last year I took in homestays and had six more visitors this year – lovely people and I made over $600. (Interview, 2009)

Winning over the majority of local people was eventually possible. A year later, a more detailed survey (Jetty Research & UTS 2010) found even more overwhelming support, recognising the short-term economic benefits at festival time and the longer-term benefits from tourism, while, significantly, at least 20% stated that their views had changed over the years, and they were now more positive about the festival. Many had participated and become involved as volunteers or additional paid staff.

Even the successful film *The Dish* (2000) had first provided something of a setback to Parkes since some saw it as making Parkes look backward by being based around the first moon landing (Brennan-Horley et al., 2007). However, its commercial success brought some new interest in Parkes and greater enthusiasm for the film. Similarly the growing success of the Elvis Festival helped to change local perceptions of the event and garner further support and interest from local businesses and the wider community. The majority of local people not just came to terms with festival, but decidedly embraced it. As one homestay host said: 'If it's good for Parkes I'll be in it'. One local man, dressed in a jumpsuit with guitar, stated:

> I've gone all out; there's no half measures in this town. I've got the wig and the suit, the rings and don't forget these awesome sunnies ... it's

tackalicious! I think if we were anywhere else we'd get bashed but around here you just get bought beers; it's fucking fantastic. (Interview, 2009)

By the mid-2000s, the mayors, and councillors, routinely dressed up and accompanied the train on the large stage to Parkes, crowds of several hundred welcomed the train, and draped visitors in leis (not only in the Hawaii years). The Festival had become part of Parkes life.

Parkes has succeeded despite the scepticism and downright opposition of some of the townsfolk, concerned about the image and status of the town. Some prefer the link to an Australian icon – 'The Dish'– as the appropriate image for a town named after the founder of Australian Federation, while others still object to what they see as a tawdry celebration of popular culture. Nonetheless, enough are well aware of the economic benefits, most stores on the main street have increasingly decorated their windows and entered into the Festival spirit, and what was once a more divided community has benefited substantially and come to terms with its strange musical identity.

Elvis has Not Left the Building

Over time, the residents of Parkes have adopted Elvis. He was never their choice as a symbol but in the end they have adapted to life with Elvis, just as the wider world has come to see Parkes as the Elvis town. For the first time in 2009, an English newspaper, *The Independent*, featured Parkes and Elvis in its travel section. In Parkes, its growing Elvis reputation led former Wiggles member Greg Page (the yellow Wiggle in the world's most popular children's television band) to choose the town to locate his collection of Elvis memorabilia (the fourth largest such collection in the world) – forming the basis of a new permanent museum. The collection was housed in its own building, next to a rather dowdy 'traditional' museum of no great interest. Parkes is now home to, amongst many other things, *the* gold lamé suit (worn by Elvis on the cover of *50,000,000 Elvis Fans Can't Be Wrong)*; as well as Elvis and Priscilla's marriage certificate, and the last Cadillac Elvis owned. The tourism potential of such a museum collection in what was previously a fairly anonymous rural Australian town cannot be overstated. Just as importantly the new museum has given Parkes a year-long Elvis presence and, in a place with no other distinct tourist attractions, provides a rationale for visiting and remaining a little longer, remembering the experience and cementing the connection between Parkes and Elvis.

Small, struggling towns in rural Australia have promoted festivals of all sorts, both as community-building exercises and because they can attract visitors. Few have been anywhere near as successful as Parkes, despite an unpromising theme and an inauspicious beginning. The Parkes Elvis Festival demonstrates how a small place can stage a festival in a relatively remote location, on a theme of no local relevance, and succeed despite itself. Indeed Parkes succeeded against long odds: droughts and floods, early local hostility and a festival site on the fringes of town, an incredulous national press and an oppressive climate. Most other festivals linked to individual musical performers try to generate a link to that performer – whether birth place, place of death or place of famous recordings (Gibson & Connell, 2005). This is not so in Parkes. Although it has now become known throughout Australia as a location associated with Elvis, Parkes has wholly invented this association.

Myth and tradition are not always tied to authenticity and credibility where tourism and festivals are concerned. Like Bundanoon (see Chapter 16), Parkes has become the site of an 'invented tradition', where a particular image has been grafted on to a place, linked to a particular imagined historic past, but assumed to have been ever present (Hobsbawm, 1983). However, unlike 'traditions' now widely if incorrectly accepted as innate (such as tartan kilts in Scotland), it is quite clear to all that there is no Elvis tradition in Parkes. The town has succeeded in spite of itself and created a celebration of kitsch, fantasy and popular culture that is as 'real' as any celebration of Elvis in Australia could be. Its many supporters derive a variety of sensory experiences and pleasures from the Festival, none more so than the serious Elvis fans for whom it is an annual ritual (Mackellar, 2009). The town has effectively, if belatedly, deployed what can be seen as 'strategic inauthenticity' (Taylor, 1997), placing the town on the tourist map, thus creating a form of 'invented geography'. Parkes' identity is no longer just as a wheat town, the home of the 'Dish' or the 'crossroads of a nation', but is also a place that resonates nostalgically of an American legend (Figure 11.6).

Parkes thus mirrors somewhat similar tourist destinations in the United States, notably Roswell (New Mexico) and Metropolis (Illinois), of which the former became the 'UFO capital of the world' (Paradis, 2002) and the latter the home of Superman, the comic book superhero who is based in the fictional city of Metropolis. In both places, festivals celebrate such invented spatial relationships. Such towns have been able to gain significant economic and social benefits by developing and trading on unlikely, improbable, even wholly fictitious and sometimes 'unworthy' events and associations. Perhaps ironically, in 2010, Parkes, anxious to develop

Figure 11.6 Parkes railway station, Elvis Revival Festival 2010 (Photo: John Connell)

Festivals at other times of the year, were contemplating a Tedfest festival for the fictitious Father Ted, the star of the cult comedy television series, simply because there was not one in Australia.

Both locals and tourists have questioned the longevity and sustainability of the Elvis Revival Festival but, so far, it has grown each year and many visitors keep coming back. By 2010, the apparent joke had become an institution supported by the local council, the state and the nation. It had also seen off competition – including competition from nearby Forbes which once mounted a jazz festival on the same weekend. In 2001, the small South Australian seaside town of Victor Harbor launched the Festival of the King, this time marking the date of Presley's death. For the first time, Parkes had a form of direct competition for the 'Elvis market', though that too proved to be short lived. Elvis had come to just one town.

Acknowledgements

We would like to thank Kelly Hendry for her continued support and assistance on our many visits to Parkes. The initial research was supported by an Australian Research Council Discovery Grant (DP0560032) and we wish to acknowledge the help of Chris Brennan-Horley, Elyse Stanes, Anna Stewart, Shane Newman, Jenny Li, Brad Ruting and Robbie Begg in data collection and analysis.

References

Brennan-Horley, C., Gibson, C. and Connell, J. (2007) The Parkes Elvis Revival Festival: Economic development and contested place identities in rural Australia. *Geographical Research* 45, 71–84.
Gibson, C. and Connell, J. (2005) *Music and Tourism*. Clevedon: Channel View Publications.
Hobsbawm, E. (1983) Introduction: Inventing traditions. In E. Hobsbawm and T. Ranger (eds) *The Invention of Tradition* (pp. 1–14). Cambridge: Cambridge University Press.
Jetty Research and University of Technology (2010) Random Telephone Survey of Residents to Measure Impact of the 2010 CountryLink Parkes Elvis Festival, UTS, Sydney.
King, C. (1994) His truth goes marching on: Elvis Presley and the pilgrimage to Graceland. In I. Reader and T. Walker (eds) *Pilgrimage in Popular Culture* (pp. 92–104). London: Macmillan.
Mackellar, J. (2009) Dabblers, fans and fanatics: Exploring behavioural segmentation at a special interest event. *Journal of Vacation Marketing* 15, 5–24.
Paradis, T. (2002) The political economy of theme development in small urban places: The case of Roswell. *Tourism Geographies* 4, 22–43.
Taylor, T. (1997) *Global Pop: World Music, World Markets*. London: Routledge.

Chapter 12
Marketing a Sustainable Rural Utopia: The Evolution of a Community Festival

M.W. ROFE and H.P-M. WINCHESTER

Reinventing Christmas

As traditional rural economies decline, new consumption-based economies are taking their place. Increasingly, these new economies are being constructed and places being re-imagined, reinvented and re-packaged to attract urban consumers. Spearheading this transition is a multitude of festivals that seek to position a given place positively within a burgeoning and highly competitive tourism marketplace. There is a proliferation of festivals spanning wine, cuisine, arts and heritage throughout rural Australia; Gibson and Connell (2005) estimate that some 600 music festivals exist in New South Wales alone. In this environment, distinctiveness provides a competitive advantage which becomes the currency of success as communities jostle for market position. The distinctiveness of place-based festivals is maintained through continued evolution to satisfy a discerning and mobile consumer market (Getz, 2002).

This study of the South Australian village of Lobethal considers the construction and diversification of its Christmas-themed Lights of Lobethal. The continued success of this festival has drawn initially on the marketing of its authentic Germanic traditions (Winchester & Rofe, 2005). The evolution has been demonstrated in two main ways, first in developing a number of sub-events which provide multiple opportunities for consumer visits. Second, the development of a theme of ecological sustainability provides a 21st-century take on the notion of the rural idyll.

Lobethal through its lights festival is presented as a rural utopia and a rural idyll. These concepts encapsulate key ideas of a place apart from the pressures of urban life, and also elements of tradition, community and security, which are woven into the Christmas theme. The diversification of the Lobethal festival into food, gardens and environment captures other elements of the rural idyll; of purity and wholesomeness, which are brought into contemporary focus by a new accent on sustainability. The combination of environmentalism and Christmas provides a reinvention of the rural idyll which brings together tradition, heritage and community with a vision of the countryside which is not only wholesome and sustaining but also sustainable. This chapter examines the development and evolution of Lobethal's festivals, set in the context of the rural utopia and rural idyll.

The Rural Idyll and Rural Utopia

In the pursuit of festival-led, consumption-oriented economies rural places are transformed, discursively at least, from spaces of production to landscapes of and for consumption. In short, new rural economies represent a significant scene shift for rural communities. This scene shift often presents the rural as a 'place apart' from the pressures of modern urban life. These notions are captured in the term rural idyll. While the emergence of new rural economies is a recent phenomenon, the origins of the rural idyll are considerably older. Indeed, narratives of the rural as a 'place apart' are well entrenched in modern societies, tracing their origins to the Industrial Revolution. Urban squalor and the fear of social decline combined to create a powerful longing for a purer and more wholesome form of existence. This longing led to the romantic construction of the rural as the preserve of community values, social stability and individual safety (see Hopkins, 1998). In this dichotomous rural/urban construction, the rural is:

> used in contrast with the fears of the present and the dread of the future ... Households can look back to rural roots ... the location of nostalgia, the setting for the simpler lives of our forebears, a people whose existence seems idyllic because they are unencumbered with the immense task of living in the present. (Short, 1991: 34)

Thus, the rural idyll has become an evocative place-making strategy in the creation of new rural economies. A burgeoning literature demonstrates how the rural idyll has been effectively mobilised to market an array of rural themed goods and services ranging from home wares and furniture to cuisine and cooking classes (see e.g. Bessière, 1998; Derrett, 2003; Panelli *et al.*, 2003; Staples, 2003). Such practice ushers in a logic where the '"[r]ural" is thus a marketed brand name for a specific kind of place commodity; the "rural" is a commodified sign of a symbolic countryside' (Hopkins, 1998: 77). Regardless of the product or the location, the rural brand draws upon the significant semiotic load of the rural idyll.

The rural brand is most powerful when embedded in a rural location. Such a place may become so imbued with the values of the rural that it is perceived as a 'heaven on earth' or a rural utopia. A rural utopia is not only a discursive marketing tool but also a physical 'reality' which is able to be experienced through interaction and ultimately consumed. Place-based rural festivals are at the spearhead of this rural branding, presenting specific places as rural utopias. Lobethal has successfully marketed itself as a Christmas wonderland through the Lights of Lobethal Festival (Rofe & Winchester, 2007; Winchester & Rofe, 2005), drawing heavily upon a carefully constructed sense of the rural brand. We contend that, at face value, Lobethal encapsulates a rural utopia.

The purpose of this chapter is to critically investigate the development and the evolution of Lobethal's rural place making, with a specific emphasis on the interweaving of different elements of the rural idyll. We are also concerned with problematising the rural brand and the ways in which the rural idyll is mobilised to create a sense of utopia for the purposes of promoting rural places for tourist consumption. In doing so, we directly respond to Newby's (1987: 1) challenge that '[t]he conventions which surround a romantic view of the countryside ... need to be cleared aside' along with the 'equally pervasive evaluation of rural life ... that nothing of importance ever happens there: that arcadian virtues exist beneath a pall of tedium'. We identify fractures in the carefully constructed identity which are both historic, relating to Lobethal's Germanic past, but also new tensions which are being thrown up in the development of a sustainable Christmas.

Lobethal: A Rural Utopia?

Located in the Mount Barker ranges some 35 km as the crow flies from Adelaide; the village of Lobethal is a small and relatively isolated rural community of 1836 people (ABS, 2007). First settled in 1842 by Germanic Lutherans fleeing religious persecution in Europe, Lobethal is one of a number of such settlements in the Adelaide Hills. From the outset, the Germanic villages were imbued with a sense of the rural idyll. Early colonial writings exemplify this, one observing that in Adelaide '[t]he weather was so hot it was almost insupportable and not a blade of grass' grew, yet

Marketing a Sustainable Rural Utopia

in the Germanic villages 'the air felt so pure and invigorating that I could not think that I was in the same country' (1840 cited in Whitelock, 2000: 360). Here the environment of the Germanic rural villages was constructed as being more pure than urban Adelaide. Aligned with nature, the Germanic communities themselves were construed to be wholesome and virtuous:

> Scarcely a day passes but that some of the [Germanic] people repair to the place of worship, wither early in the morning, or in the evening after work ... There is nothing more pleasing for the passer-by than to have the voices mingling in sacred song rise and fall upon the ear. (cited in Whitelock, 1985: 256)

Indeed, the Germanic Lutheran settlers were very religious. Consequently, the Germanic Lutherans were viewed as ideal settlers in a colony designed to be a 'wealthy, civilized society' (Wakefield cited in Whitelock, 2000: 3) free from convict transportation. Such depictions constitute the origins of the construction of the Germanic Lutheran villages as idyllic rural landscapes. In the case of Lobethal, these depictions are reinforced by contemporary discourses that promote the village as a tourist destination hosting several festivals generally and as a Christmas wonderland specifically. We now turn to a discussion of these.

Today, Lobethal is marketed as 'the kind of place where people still make eye contact and smile when they pass you in the street' (Lobethal Community Association, 2009: n.p.). Community websites, newspaper articles and other marketing materials espouse Lobethal as a place untouched by the pressures of modern urban life. Indeed, in promoting the village as a tourist destination, even the journey to Lobethal is constructed as a cathartic experience:

> Travelling to Lobethal along rambling country roads winding through vineyards and orchards you leave the stress of city life behind. The air is crisp and clean. The township of Lobethal is situated in a valley which enhances its village feel. (Lobethal Community Association, 2009: n.p.)

This journey is both physical and symbolic. The act of leaving the city, of rambling through a picturesque countryside to an idyllic village is implicitly transformative. These sentiments echo Hopkins' (1998) claim that a core aspect of the rural idyll is alterity. Rural alterity refers to the way the 'rural is represented as being some place other than urban, as some time other than the present, as some experience other than the norm'

(Hopkins, 1998: 78). Lobethal's carefully constructed sense of the rural idyll positions the village and the experiences it offers as being beyond the urban and apart from the modern word. Pragmatically, Lobethal's proximity to Adelaide enhances the ease with which this cathartic journey can be taken. As Taylor and Shanka (2002) note, places located within day-trip proximity to major urban areas are well positioned to attract high tourist numbers improving the success of festivals (see also Chapter 1). Lobethal is well positioned geographically, being close to Adelaide, yet retaining its rural landscape due to its relative isolation.

Lobethal's construction as a rural utopia is carefully managed and multifaceted. In the marketing of the village senses of a romanticised past are carefully entwined with a contemporary, cosmopolitan present, albeit always rural branded and imbued with a sense of understated authenticity:

> In and nearby Lobethal you will find heritage buildings; local artists; fresh, luscious apples and pears; wine tasting; local seasonal produce; boutique beers; eclectic market stalls; motorbikes; polocrosse horses and alpacas. Why not treat yourself to a day in the country. See our Visitor Information section for further information about the understated delights of Lobethal. (Lobethal Community Association, 2009: n.p.)

Lobethal offers tourists a broad range of experiences to attract a diverse array of consumers. Spearheading these experiences is a number of major festivals, addressing the heritage, cuisine and community market segments.

The Lights of Lobethal Festival

While, as discussed below, a series of other festivals and events are successful and highly important aspects of Lobethal's rural brand, the village is most renowned for its annual Lights of Lobethal Christmas festival. This festival transforms Lobethal into a Christmas wonderland. Held over a 17-day period, this festival attracts over 250,000 visitors. Central to the festival is a carefully constructed intersection between the village's Germanic Lutheran heritage and romanticised notions of community. This intersection draws heavily upon the rural idyll's themes of tranquillity, kinship and community. During the festival, this discourse becomes manifest through Lobethal as a physical place transforming the village's landscape.

During the Victorian Period, Christmas was construed to be emblematic of the rural idyll (Connelly, 1999; Miller, 1993; Winchester & Rofe,

2005). Depictions of traditional Christmas celebrations from this period unfailingly locate them within idyllic rural settings:

> the only way ... to see Christmas ... surrounded by all its poetical associations, is to spend it in the country. In town it is tricked out in the new fashions – very pretty to look at, yet in nowise romantic. But there are ... *out-of-the-way nooks ... which, lying from off the great high roads, seem to have been forgotten by ... time ... These are the spots where you feel the Poetry of Christmas to its full.* (Illustrated in London News, 24th December, 1853, cited in Connelly, 1999: 27) (Emphases added)

The above excerpt articulates the rural/urban dichotomy most clearly. Further, the assertion that the true 'Poetry of Christmas' can only be experienced to it fullest in rural places that themselves have been 'forgotten by time' locates the rural brand in both time and space. The Lights of Lobethal explicitly mobilises these notions, presenting a *traditional* community event that is freely given as 'a gift from the people of Lobethal to the wider community' (promotional brochure, 2003, cited in Winchester & Rofe, 2005: 273). Interviews with festival visitors conducted by Winchester and Rofe (2005: 273) reveal the impact of this gift:

> I'm quite surprised that a town would open itself like this. I guess this is a part of their ... Christmas culture. *I think that they must have great pride in their own community and knowing that they can give peace to the wider community.* I think it is a good aim and gives the community solidarity ... there is a sense of openness. (Winchester & Rofe, 2005: 273) (Emphases added)

This 'sense of openness' is in direct contrast to perceptions of the urban. Further, the reference to the welcoming and generous nature of the community harkens after the nostalgic longings for more communal forms of life so central to the rural idyll. As a gift, the festival is not simply about visiting, but participating.

Diversifying Christmas

The organising committee has progressively diversified the festival to include a significant number of Christmas-themed events, including an opening firework and 'lights up' event, a nightly live nativity play, a pageant, street parade and Christmas tree display. New for 2009, an environmentally friendly lighting program called Deck the Hall (see Figure 12.1) and solar lighting displays and workshops and a carbon neutral forum were introduced. Deck the Hall and associated sustainability themed

Figure 12.1 Deck the Hall, Lobethal (*Source*: http://lightsoflobethal.com.au/lol/deckthehall.html 2009)

events reflect the need for places to diversify their place-making strategies in order to remain vital in an increasingly competitive market. These initiatives represent a significant evolution in the festival specifically and the marketing of Lobethal generally. Through these initiatives Lobethal's identity is evolving into a sustainable eco-village. Here, 18th-century notions of rural wholesomeness and environmental purity evolve into a 21st-century discourse of sustainability. The marketing for Deck the Hall and its related Solar Grove Workshop, exemplify this:

> Lobethal is one of the secret yet beautiful hidden treasures of the Adelaide Hills, and the Lights of Lobethal is a very special time of year, when the local community turn[s] it into an illuminated paradise for visitors to enjoy and celebrate Christmas time. It's an incredible event because the whole event is created by community participation ... with hundreds of homes being lit up by local businesses and families as a community gift to visitors from everywhere. It is quite unlike other festivals. (Anon., 2009a: n.p.)

The above is replete with expressions of the rural idyll that Lobethal and other rural communities have normalised as an inherent aspect of their landscapes and identities. These can be considered as the 'gifts' of

Marketing a Sustainable Rural Utopia 201

the rural. However, to these established gifts, Lobethal is inserting that of sustainability:

> The many gifts of Lobethal ... include: the beautiful Lights of Lobethal Festival and how it brings people together; a close community; *amazing trails for walking and cycling*; a great place to grow up; and strong heritage. (Anon., 2009a: n.p.)

New sustainability initiatives, such as the Solar Grove Workshops and Carbon Neutral Forum, represent a significant development of the festival (see Figure 12.2). These events were marketed to local residents '... thinking about global warming' but who may not have an awareness or understanding of these '... new industries and new values plus a lot of new [environmentally-oriented] words that have come into our language' (Anon., 2009b: n.p.). While these events were explicitly premised on contemporary action with a future focus they were also imbued with a sense of nostalgic country home-craft; '[h]ave you been wondering about new technology ... but you enjoy a bit of old fashioned craft?' (Anon., 2009c: n.p.).

Figure 12.2 Solar Grove, Lobethal (*Source*: http://lightsoflobethal.com.au/lol/solargrove.html 2009)

All of these sub-events encourage visitors to move through the village, enjoying different aspects of the festival. In this way, the festival has, for some visitors, shifted from being a spectacle to visit at Christmas to being integral to their own family Christmas traditions. Visitors' books at the spectacularly decorated home themed as Santa's Retreat record the positive impact of this festival and its incorporation into Christmas traditions beyond the host community; 'we love it – it is part of our family Christmas tradition to come to Lobethal every year' (cited in Winchester & Rofe, 2005: 274). Scrutinising why festivals fail, Getz (2002: 214–215) proposes that competition from similar festivals and a failure to evolve leads to visitor numbers declining. Given the long history of the Lights of Lobethal and its continued evolution, as evidenced by new and innovative projects such as Deck the Hall, it is foreseeable that the festival will remain vibrant and viable into the future. A more important factor ensuring the longevity of the festival may well be its incorporation into the Christmas traditions of families from beyond Lobethal.

Diversifying the Rural Idyll

While the village is renowned for its lights festival, other festivals are held throughout the year; in 2009, including the Hills Market Kitchen for Kids, Food Affairs with Flair, Lobethal Grand Carnival and the Hills Garden and Environmental Expo. These events attract significant local, national and international attention. The Grand Carnival, for example, celebrates Australia's first motor racing grand prix held between 1939 and 1948. In 2009, the Grand Carnival celebrated the 70th anniversary of the inaugural race, attracting some 110 vintage racing cars and bikes worth an estimated A$60 million (Martin, 2009). It could be asserted that the Grand Carnival is not directly aligned with the tranquillity typically associated with the rural idyll. However, a close reading of the Grand Carnival's marketing reveals a deliberate intersection between two key facets of the rural brand, namely the rural as a space of living heritage and rural places as spaces of freedom and decisive action as opposed to the controlled environments of the urban:

> For students of Australia's motor racing past, for the handful of surviving drivers and riders, and for those who were there, as crew, as spectators, as awestruck youngsters; the word 'Lobethal' is at once synonymous with magnificent racing machines, daunting speeds on a sealed and hugely challenging circuit of more than eight miles, and heroic measures of human bravery, inspired performance and epic

endurance. Nestled in the Adelaide Hills, less than an hour east of Adelaide, Lobethal is set for a serious re-enactment of its glory days. (Heaps Good, 2009: n.p.)

Lobethal's diversity of festivals reflects an astute awareness among the community that economic success cannot be premised upon a single event and the market image that this represents. This awareness and the internationalisation of Lobethal's rural brand, affirms Bessière's (1998: 22) observation that '[r]ural populations have extended their networks, widening their social space and economic scope' through deliberate place marketing strategies. This widening necessarily involves an acute awareness of local 'image and of the importance of local culture' (Ekman, 1999: 282) and how these may be effectively harnessed.

Fracturing Lobethal's Idyllic Construction

Beneath Lobethal's carefully constructed identity as a traditional, wholesome rural village generally, and a Christmas wonderland specifically, a number of fractures and collisions lurk. Rather than a history of community harmony and a valued contribution to South Australia, Lobethal's history is in fact riddled with religious schisms, hardship, xenophobic discrimination and economic upheavals including the 1993 closure of the Onkaparinga Woollen Company with the loss of some 400 jobs (Rofe & Winchester, 2007; Winchester & Rofe, 2005). These accounts are absent from the marketing of the village as they unsettle the notion of Lobethal as a harmonious, idyllic place. Historically, the depth of anti-German sentiments within South Australia during World War I is exemplified by the Government's Nomenclature Committee which investigated instances of 'enemy' place names within the State. Ultimately, 69 place names of Germanic origin were expunged and replaced with more suitably English or 'native' Australian names. In 1917, Lobethal became Tweedvale and remained so until 1935. The Lutheran day school was also closed as were German language newspapers. Ironically, the Roll of Honour in Lobethal's Centennial Hall which honours the village's young men who died in this war reveals that the majority of their family names are of Germanic origin. This episode rates only a passing mention, if any, on tourist websites promoting Lobethal. While not unsurprising, it reinforces that 'the past is not simply given. It is the arena for struggles over remembering and forgetting, as people try to claim the future' (Metcalfe & Bern, 1994: 665).

Within the Lights of Lobethal Festival, a number of fractures are identifiable which problematise the carefully constructed sense of Lobethal's rural brand. These are fractures between the secular and the sacred, community versus commerce, harmony versus disruption and inclusion versus exclusion (Rofe & Winchester, 2007; Winchester & Rofe, 2005). Some of the fractures are often subtle and beyond the direct experience of the visitor. An example of an invisible difference is the potentially exclusionary nature of such an overtly Christian landscape in a pluralistic Australia (Winchester & Rofe, 2005: 276–277). A visually striking difference is the appearance of banners offering competing visions of the spirit of the season, as CHRISTmas and Christmas (Rofe & Winchester, 2007: 146–147; see Figure 12.3). Others are also readily apparent, such as notices warning visitors of the potential for petty crime (see Figure 12.4) or the traffic congestion that is commonplace.

The evolution of the Christmas festival into a more diverse range of events offers the potential for more divisions and fractures. Some of the 21st-century developments sit more easily with the rural utopia than others. The development of an environmentally friendly rural brand with

Figure 12.3 Celebrating CHRISTmas as opposed to Christmas. (*Source*: Winchester, H.P.M and Rofe, M.W. (2005) Christmas in the 'Valley of Praise': Intersections of the rural idyll, heritage and community in Lobethal, South Australia. *Journal of Rural Studies* 21, 265–279)

Marketing a Sustainable Rural Utopia 205

Figure 12.4 Welcome to Santa's Retreat … Lock Your Car! (*Source*: Winchester, H.P.M. and Rofe, M.W. (2005) Christmas in the 'Valley of Praise': Intersections of the rural idyll, heritage and community in Lobethal, South Australia. *Journal of Rural Studies* 21, 265–279)

solar lighting resonates with historic views of Lobethal outlined earlier. The ecological sustainability of the solar lighting adds a contemporary edge to the utopian argument, which rounds out the discourse of community and heritage implicit in the Christmas story. In practice, however, the juxtaposition of walking and cycling with the congested vehicular traffic trails is likely to be difficult to manage on narrow country lanes. A potentially more serious fracture in the rural utopia is presented by the Grand Carnival of vintage racing. While this event draws on a discourse of generic heritage, it lacks the specific Germanic influence and also represents a severe disjuncture with the desired aim of ecological sustainability. The Carnival, which is being deferred for 2010/2011 'while changes to the event format and financial structure are reviewed' (Lobethal Grand Carnival, 2010: n.p.), is represented as appropriate for rural spaces as a theatre for freedom and decisive action. However, it remains to be seen how these multifaceted aspects of rurality can be drawn together in a coherent image or even whether such coherence is desirable or necessary.

Conclusion

This chapter has sought to critically engage with the complexities of rural place making and to explore how one South Australian community has embraced the opportunities presented by a new, festival-led economic base. Typically, the pursuit of these opportunities in rural areas is associated with some form of commodification of the rural idyll. Emerging during the Industrial Revolution as a reaction to urban squalor and social decline, the rural idyll is a powerful discourse deeply embedded in the Western psyche. Central to the rural idyll are notions of the countryside as the natural domain of wholesome values such as community and kinship and a landscape of religious piety and social stability. These discourses are embedded within contemporary rural branding that presents specific places as rural utopias. The village of Lobethal is an exemplar of these processes. Lobethal's construction as an idyllic rural retreat from the adjacent city of Adelaide draws carefully upon romanticised notions of its Germanic Lutheran heritage. Specifically, the village's Christmas-themed Lights of Lobethal festival transforms the village into an Arcadian wonderland that is presented as a selfless gift to the wider community. While this gift draws upon the village's heritage and is constructed as an integral aspect of the community's fabric, recently its scope has been expanded to include not just notions of safety and selflessness, but also sustainability. Here, the wholesome qualities of the rural as an idyllic 'place apart' from the urban are evolving into a message of sustainability. Lobethal's identity and landscape are increasingly presented as being both sustaining and sustainable. Indicative of this is the development of the Lights of Lobethal Festival to include workshops and seminars on sustainable concepts and practices. These innovations can be taken as local recognition of the need for the constant reinvention of existing festivals so as not to lose their market appeal in an already crowded and highly competitive tourism-driven marketplace. Further, they can also be read as a grass-roots' response to environmental issues beyond the scale of local. This evolution represents both a point of market difference for the festival organising committee and an interesting refashioning of the rural idyll discourse projecting it from the 18th into the 21st century.

Despite the obvious success the village enjoys, a close reading of the Lights of Lobethal and other events reveals a number of fractures that problematise Lobethal's idyllic rural brand. While none of these are construed to be, at this stage, particularly significant, they render visible the constructed nature of rural place making. These fractures and tensions aside, the Lights of Lobethal festival is a resounding success and, in conjunction with the other festivals hosted by the community, has brought significant economic benefits to the village. In this sense, Lobethal has been fortunate in that it has been able to draw upon positive, even romantic notions of events that encapsulate well the essence of the rural idyll, despite periods of turmoil in its past. Thus, it has been a relatively easy exercise for Lobethal to present itself as a rural utopia. The challenge for the future will be to manage ongoing and competing discourses as it continues to develop and reinvent its events and activities.

References

Anon. (2009a) The many Gifts of Lobethal ... discovering them anew. On WWW at http://www.illuminart.com.au/lobethal/archives/date/2009/11. Accessed 06.06.10.

Anon. (2009b) Carbon neutral ... which gear is that? Where do I get it? What is it made of? On WWW at http://www.illuminart.com.au/lobethal/archives/date/2009/11. Accessed 06.06.10.

Anon. (2009c) The Solar Grove Workshops for Lobethal residents. On WWW at http://www.illuminart.com.au/lobethal/archives/date/2009/11. Accessed 06.06.10.

Australian Bureau of Statistics (ABS) (2007) *Census of Population and Housing Lobethal (UCL 412400) Basic Community Profile*. Canberra: Australian Government Publishing Services.

Bessière, J. (1998) Local development and heritage: Traditional food and cuisine as tourist attractions in rural areas. *Sociologia Ruralis* 38, 21–34.

Connelly, M. (1999) *Christmas: A Social History*. London: I.B. Tauris Publishers.

Derrett, R. (2003) Festivals and regional destinations: How festivals demonstrate a sense of community and place. *Rural Society* 13, 35–53.

Ekman, A. (1999) The revival of cultural celebrations in regional Sweden. Aspects of tradition and transition. *Sociologia Ruralis* 39, 280–293.

Getz, D. (2002) Why festivals fail. *Event Management* 7, 209–219.

Gibson, C. and Connell, J. (2005) *Music and Tourism: On the Road Again*. Clevedon: Channel View Publications.

Heaps Good (2009) Lobethal Grand Carnival. On WWW at http://www.heapsgoodsa.com.au/Item/Detail.aspx?p=19&item=498. Accessed 06.06.10.

Hopkins, J. (1998) Signs of the post-rural: Marketing myths of a symbolic countryside. *Geografiska Annaler B* 80, 65–81.

Lobethal Community Association (2009) The Community of Lobethal. On WWW at http://www.lobethal.sa.au/. Accessed 06.06.10.

Lobethal Grand Carnival (2009) Event deferred until 2012. On WWW at http://www.lobethalgrandcarnival.com.au/home.html. Accessed 06.06.10.

Martin, S. (2009) Hills blast from the past. *The Advertiser* (October 2nd). On WWW at http://www.news.com.au/adelaidenow/story/0,26152346-2682,00.html. Accessed 06.06.10.

Metcalfe, A.W. and Bern, J. (1994) Stories of crisis: Restructuring Australian industry and renewing the past. *International Journal of Urban and Regional Research* 18, 658–672.

Miller, D. (ed.) (1993) *Unwrapping Christmas*. Oxford: Clarendon Press.

Newby, H. (1987) *Country Life: A Social History of Rural England*. Totowa, NJ: Barnes and Noble.

Panelli, R., Stolte, O. and Bedford, R. (2003) The reinvention of Tirau: Landscape as a record of changing economy and culture. *Sociologia Ruralis* 43, 379–400.

Rofe, M.W. and Winchester, H.P.M. (2007) Lobethal the Valley of Praise: Inventing tradition for the purposes of place making in Rural South Australia. In R. Jones and B.J. Shaw (eds) *Loving a Sunburned Country? Geographies of Australian Heritage* (pp. 133–150). Aldershot: Ashgate.

Short, J. (1991) *Imagined Country: Society, Culture and Environment*. London: Routledge.

Staples, M. (2003) Tasmania as Little England and the social construction of landscape. *Rural Society* 13, 312–328.

Taylor, R. and Shanka, T. (2002) Attributes for staging successful wine festivals. *Event Management* 7, 165–175.

Whitelock, D. (1985) *Adelaide: From Colony to Jubilee, A Sense of Difference* (2nd edn). Adelaide: Savvas Publishing.

Whitelock, D. (2000) *Adelaide: From Colony to Jubilee, A Sense of Difference* (3rd edn). Adelaide: Arcadia Press.

Winchester, H.P.M and Rofe, M.W. (2005) Christmas in the 'Valley of Praise': Intersections of the rural idyll, heritage and community in Lobethal, South Australia. *Journal of Rural Studies* 21, 265–279.

Chapter 13
ChillOut: A Festival 'Out' in the Country

G. WAITT and A. GORMAN-MURRAY

Introduction

Little is known about how sexual diversity shapes lives outside metropolitan Australia. Places beyond the metropolis are conventionally seen as fashioned by narrow strictures of heterosexuality (Gottschalk & Newton, 2003). A host of (post)colonial national mythologies encourage people to imagine rural Australians as white, heterosexual men and women. Such ideas are encapsulated in the bushman mythology and the legend of Ned Kelly, and replayed through characters such as Crocodile Dundee (Turner, 1994), as well as the figure of the 'farm woman' and long-standing organisations such as the Country Women's Association (Alston, 1995). Where sexual diversity enters this imaginary, visions are often overlain with narratives of lesbian and gay suicide, homophobic violence and 'escape' migration to 'gay' urban centres. This chapter presents a counternarrative, offering stories about the enjoyment of ChillOut, a lesbian and gay festival held annually in Daylesford-Hepburn Springs, twin country towns in Hepburn Shire, Victoria. This is partly an exploration of how ChillOut complicates and challenges the 'closeting' of lesbians and gay men in rural Australia, utilising camp humour to unsettle stereotypical assumptions about (people of) diverse sexualities. But at the same time, ChillOut is influenced by economic discourses that frame lesbian and gay tourists as 'affluent', and is embedded in political structures that continue to favour heterosexuality as 'normal'. In this light, the festival must also be considered through the construction of the 'gay tourist' or 'lesbian tourist' as subjects of both privilege and opposition. Drawing on these perspectives, we address the different experiences and meanings of ChillOut for organisers, participants and residents.

To explain how Hepburn Shire has become a festival 'capital', this chapter first outlines its shifting demography, economics and politics. We pay particular attention to the origins and development of ChillOut. We then explore how the playfulness of camp unsettles the assumptions of rural places devoid of sexual difference. Next, we argue that while ChillOut challenges the absence of diverse sexualities in many rural Australian narratives, it normalises non-threatening expressions – certain configurations of gayness as affluent lesbian and gay tourists rather than an overarching acceptance of sexual difference in all aspects of everyday life. This may be helpfully thought about in terms of a hierarchy of homosexualities, highlighting ongoing oppression for anyone whose sexuality does not conform to idealised notions of the lesbian or gay tourist. In the conclusion, we reflect upon how ChillOut tests, adapts and recreates narratives about what it means to live in Daylesford–Hepburn Springs.

Daylesford–Hepburn Springs: A Historical Social Geography

The twin country towns of Daylesford–Hepburn Springs have a population of around 3500 people. Located in the Central Highlands of Victoria, 100 km north-west of Melbourne, Daylesford is the administrative centre of Hepburn Shire. Founded by British and European settlers during the Victorian gold rushes of the 1850s, many from the Swiss-Italian canton of Ticino, place names are now the only traces of mining. At the end of the 19th century, Daylesford–Hepburn Springs was overlain with a variant of the European rural idyll-based 'spa tourism', enabled by new modes of rail mobility (Gervasoni, 2005). This brought prosperity in the early 20th century. By mid-century, however, the spas had lost their appeal for Melbourne's bourgeoisie, and Hepburn Shire was economically and socially marginalised. Businesses lay empty, the railway became disused. From the 1960s onwards, the social composition was significantly altered by the arrival of new waves of migrants from Melbourne, particularly those who did not conform to normative ideas of white hetero-nuclear families, such as 'hippies', single mothers, lesbians and gay men (Mulligan et al., 2004). Marginalised from mainstream society, these non-conventional families were attracted in part by the low costs of housing. Moreover, they too were drawn by variants of the rural idyll, imagining rurality as natural, aesthetic, therapeutic and spiritual. Daylesford–Hepburn Springs became home to many social movements challenging accepted 'mainstream' norms – heteropatriarchy, monogamy and consumerism. In the

1980s, for instance, ConFest was held at nearby Glenlyon, championing the values, practices and desires of a proposed 'New Age'.

Since the 1990s, the key economic sectors of Daylesford–Hepburn Springs shifted to service activities supporting tourism – comprising 18.6% of the workforce at the 2001 Census – rather than those centred on primary production, such as potato farming, the abattoirs and timber milling – only 9.7% of the workforce at the 2001 Census. This transformation can in part be attributed to the economic growth of Melbourne (especially during the first decade of the 21st century), increased car ownership, and the emerging desires of urban residents for a European rural idyll framed as aesthetic, spiritual, restful, relaxing and rejuvenating. Tourism values attached to Daylesford–Hepburn Springs consequently reiterate those of the late 1800s when the towns were first pitched as 'Australia's Spa Country'. Hence, tourism in Daylesford–Hepburn Springs makes few references to Australia's bush mythologies.

Given the township is now only 90 minutes' drive from Melbourne, the European rural idyll has widespread appeal, and tourism boomed. Even by the mid-1990s, an estimated 2500 people arrived every weekend. Some people decided to buy weekend homes for 'rural getaways'. Property prices soared and rental accommodation became scarcer and more expensive.

Moreover, the retail geography of the town centre has also radically changed. Restaurants, cafes, massage parlours and beauty centres announce Daylesford–Hepburn Spring's resurgence as a centre for a decadent weekend 'escape' for Melbournians (Figure 13.1). These social and physical changes inform narratives of tourism 'destroying' the identity of the town. Public forums and debates quickly polarised along the lines of 'anti' and 'pro' tourism, and little attention was given to ambiguous outcomes.

Festivity and Celebration in Daylesford-Hepburn Springs

In the last two decades, the socio-economic changes and challenges of Daylesford–Hepburn Springs were accompanied by an increase in festivals, such as the Boite Singer's Festival (1990–), the Swiss-Italian Festa (1993–) and the Words in Winter Celebrations (2003–). The growth of festivity in Daylesford–Hepburn Springs appears to be a response to an absence of events that made sense of the increasing diversity present in 'everyday' Daylesford–Hepburn Springs (van Gennep, 1960). Festivity as a response to social change is well documented. In various geographical and historical contexts of social upheaval, people organise festivals to help make sense of their lives. The festival flurry in Daylesford–Hepburn Springs mirrors what

Figure 13.1 Vincent Street, Daylesford (Photo: Gordon Waitt)

Picard (2006: 48) termed 'cycles of festive re-enchantment', where festivity enables possibilities to express place attachments during uncertain times. Voluntary organising committees in Daylesford–Hepburn Springs began publicly celebrating local sports, farming equipment, craftwork, words and writers, food and wine, flowers, singers, quilters and migrant heritage (particularly the history of immigration to Victoria from Scotland and to Daylesford from the Swiss-Italian canton of Ticino).

Within this context, in 1997, a network of lesbian and gay business owners, Springs Connections, organised a family picnic day attended by around 30 people. Since 1997, ChillOut has been gradually re-branded by the organising committee to appeal to a wider regional and metropolitan audience. The inaugural 1997 festival was a gathering of 50–100 family and friends; by 2001, ChillOut was being pitched both to 'rural visitors', with the claim that the 'event ... unites not only the gay community, but the country community as a whole', and also visiting 'city dwellers', who were invited 'to soak up all that spa-country has to offer in a relaxed and fun atmosphere, where all gay and lesbian people, their friends and families can gather and enjoy bands, games and stalls' (Anon., 2001: 7). Over 5000 people attended the festival that year. The following year, 8000–9000 people came and ChillOut was described in the Hepburn Shire press as 'Australia's largest rural gay and lesbian event' (Anon, 2002: 28). The organisers now use this superlative in pitching ChillOut to visitors on the festival's official website, and the number of attendees has continued to grow. When we conducted our research, in 2006, 16,000 people attended, and according to the official website, 25,000 people went to the 2010 festival. In transforming ChillOut from a gathering of friends to a tourist event, the organising committee paid close attention to safety, marketing, entertainment and economics, so that the festival is now enmeshed in both the state apparatus and market consumption.

ChillOut is sanctioned by Hepburn Shire Council. Although a non-profit event, ChillOut is strategically marketed in the Melbourne and Sydney lesbian and gay press as part of the summer 'party' circuit. An anonymous article from *Bent* (Anon., 2006: 67), a freely distributed lesbian and gay publication in Sydney claimed that the 2006 ChillOut:

> weekend boasts more than one would expect of a [Sydney Gay and Lesbian] Mardi Gras with big name acts lined up ... And the biggest and best way to finish up a weekend of playing in the bush is naturally – a massive dance party, complete with some of the best and diverse DJs Australia has to offer. (Anon., 2006: 67)

Play and pleasure are promoted by ChillOut. The discourse used by organisers is of 'clients', 'attendees', 'parades' and 'parties'. They never use terms such as 'protestor', 'march' or 'rally'. This is not surprising, because conviviality was always its intent. Yet, symbols of Pride play an important role in staging this tourist space: businesses are encouraged to display rainbow flags, purchased as a fund-raising activity – a reminder of how ChillOut is entangled in volunteerism, identity politics, fund-raising and market consumerism.

As a lesbian and gay tourist event, timing, ticketing and programming are crucial. The festival is held at the end of summer over the Victorian Labour Day Long Weekend, around two weeks after the Sydney Gay and Lesbian Mardi Gras and a month after Melbourne's Midsumma Festival. Each year, event organisers incorporate more free and ticketed events. Since 2003, the program has included nationally acclaimed entertainers. Extending over four days, the 2006 program offered free events such as bush walking and a street parade, and events-at-cost, including horse-riding, a golf tournament, drag shows, the Bob Downe dinner and show, a carnival day and a dance party. Significant attention was given to setting various staged events to assure attendees receive value for money, with little risk of exposure to homophobia. One major United States-based lesbian and gay website (www.gay.com) rated ChillOut more enjoyable and more important than the Sydney Gay and Lesbian Mardi Gras, largely because of its setting, which provides a point of difference from most lesbian and gay festival events (see the official website: http://chilloutfestival.com.au/History_Purpose_Aims). Pleasure seems almost assured for visitors attending this nationally and internationally profiled rural tourist attraction.

But, in simultaneously seeking to attract Hepburn Shire residents to the event, while play and pleasure of the festival remain paramount, primacy is given to discourses of charity rather than tourism. Public support is mobilised by appealing to shared values of volunteerism. Fund-raising opportunities are created for nearby social groups that are mainstream bastions. For instance, at ChillOut, Carnival Day stalls are run by the Rotary Club, football clubs and primary schools. Equally, revenue generated from entry tickets to Carnival Day are a fund-raising activity for Hepburn Shire services, including health care, aged care facilities, schools, the bushfire service and the Daylesford Hospital. These connections between often disparate social groups offer a way to normalise sexual difference for residents of Hepburn Shire.

Camp Symbols and Performances at the ChillOut Street Parade and Carnival Day

Camp – to act in an effeminate, exaggerated or parodying manner – is a well-known attribute of lesbian and gay culture, and is both an outcome and a defiance of oppression. Lesbian and gay festivals are adept at employing camp (Johnston, 2005). Camp plays on the awareness that there is no essential truth about gender and sexuality; as Probyn (1993: 505) reminds us, 'one of camp's most obvious and serious pleasures is precisely the way in which any truth becomes yet another conceit to be played with'. Camp calls into question how certain versions of sexuality are legitimate, while others are unacceptable.

Camp sensibility is central to the themes and activities of ChillOut. In 2006, the festival adopted the title 'A Decade of Fairy Tales' – a theme which enabled camp fairy tale performances at the Street Parade, held along Vincent Street on Sunday morning, to critically engage with heteronormative notions of family, romance and citizenship. For instance, one entrant chose to parody the traditional narrative of the fairy tale genre, telling an alternative story with a float comprised of four adults and seven young children – with fairy wings and blue wigs – sitting on bales of hay (Figure 13.2). The tractor pulling the float was driven by a mature-aged man, dressed as a stereotypical 'old farmer', as depicted in many fairy

Figure 13.2 Camp 'fairy tale' float, ChillOut 2006 (Photo: Gordon Waitt)

tales (and the driver may well have been an old farmer!). A sign read: 'Once upon a time there was an old farmer who had four beautiful queens and seven little princesses, who all lived together happily, in Daylesford. The End!' This subtly reconfigured fairy tale disrupted the spectators' ideas of conventional family narratives, undermining the rigidity of socially constructed heterosexual norms.

Additionally, the Street Parade included performances from national and international drag queens. As Butler (1990) and others like Probyn (1993) have discussed, drag queens epitomise camp embodiment through their exaggeration of femininity. Along Vincent Street, the drag queens subverted the assumed naturalness of gender in Daylesford–Hepburn Springs by adorning sparkly low-cut dresses, fashioning big hair and wearing make-up that overemphasised their eyes and lips. In further displays of excess, they blew kisses and lip-synched 'gay anthems' from the backseats of open-topped cars. While being an ironic colonisation of Vincent Street by camp performances typically associated with 'drag clubs' in 'metropolitan' gay communities, such displays were also integral for selling ChillOut as a gay tourism event.

Similarly, camp was crucial to the staging of the Carnival Day in the Victoria Park showgrounds (Figure 13.3). This event was clearly demarcated by a fence, with entry restricted by ticket. The Carnival Day entertainment focused on a main stage, a central arena and over 300 stalls, ranging from fundraising for primary schools to environmental groups and sexual health. Around the main stage, camped-up notions of rurality were provided by straw-bale seating, a podium comprised of two oversize trailers, and a bucking-bronco in the shape of a giant penis. Camped-up notions of femininity and masculinity were also provided by the cabaret performer and host, Dolly Diamond, and musical entertainment from 100% Kylie (a cover act of Australian 'gay diva' Kylie Minogue) and Queen, 'It's a Kinda Magic' (a cover act of legendary British band Queen, whose lead singer, Freddie Mercury, was an iconic gay performer). The staging also encouraged exaggerated performances of gender and sexuality for many people entering this space. Particularly prominent was the obvious pleasure from camp performances and parodies of Western frontier masculinity – riding a 'bucking penis', displays of pumped muscle and pink cowboy hats. This is doubly ironic, since it not only queers stereotypical images of rural masculinity,

Figure 13.3 Carnival Day, ChillOut 2006 (Photo: Gordon Waitt)

but does so by camping up a foreign 'country' style borrowed from North American frontier mythologies (rather than Australian imaginaries of the bushman). Having demonstrated the role of camp in the sexual politics of pleasure at the festival, we now explore the wider social meanings of challenging naturalised assumptions surrounding heterosexuality.

Conflicting Narratives of Lesbian and Gay Sexuality in Daylesford-Hepburn Springs

Johnston (2005) terms the rupturing of heterosexuality by lesbian and gay festivals as 'deconstructive spatial tactics', while Munt (1995: 134) argues such transformations produce a 'politics of *dis*location'. For instance, a high-profile, openly gay business person and member of the organising committee confirmed the role of camp in redefining understandings of 'gay':

> It's good sending ourselves up. The locals get to know the person, laugh and joke. That has broken down the barriers over the years. Locals think that being gay is not so bad, that gay people can be honest. The festival probably creates the interactions that enable this to happen. Local people eventually say, 'Stuff it, okay, I am going in. If I go with a mate then I will be okay'. (Interview,[1] 2006)

This key informant invokes self-mockery and being able to joke about people's reactions as essential to undermining hegemonic assumptions that position gayness as immoral, untrustworthy and predatory. New connections are made possible within the festival timeframe. Meanwhile, Natasha, a local lesbian resident, recalled how camp created an opportunity for carnivalesque transgressions amongst a local social group once notorious for their homophobia, members of an Australian football club:

> Some footies [male playing members of an Australian football club] came in towards the end to check out what was happening at ChillOut. They wanted to have their photograph taken with Kylie [the cover act of Kylie Minogue, Australian pop diva and gay icon]. It was amazing to see those straight guys posing with Kylie. Another year we had Miss and Mr ChillOut. Mr ChillOut was this footie guy. He was prancing around in his shorts, while the crowd was screaming. It was fun! (Interview, 2006)

Similarly, Lucy, a heterosexual single mother and community worker in her 50s, who had lived in Daylesford for over 20 years, spoke about how the growth of the festival has helped reconfigure understandings of Daylesford–Hepburn Springs:

> I think it has had a positive impact. When it first started, the first parade we drove into, cos it was going to start at 10 ... by the time we parked our car it was over, it was so short, so every year it's got bigger. It's always fun, it's always a celebration and straight people are very involved in it. In fact, I mean, they will invite straight people to be involved ... So yeah, it is a weekend of, if you don't accept gays and lesbians well, it's *you* that's got to stay home!

Lucy highlights how normative conventions are overturned during the ChillOut weekend. Those residents who are least accepting of sexual diversity may choose to stay away. Since ChillOut is licensed by Hepburn Council, the social implications could be interpreted to provide what Markwell (2002: 90) terms 'gay times', a temporary time–space sanctioning social transgression.

However, various residents we interviewed suggested that the social impacts of ChillOut are more spatially and temporally diffuse. Indeed, several heterosexual residents described Daylesford–Hepburn Springs as a 'gay capital'. Most affirmatively, Emma asserted she felt pride in disclosing this aspect of Daylesford–Hepburn Springs's identity:

> I feel personally very proud. I always tell people we're the gay capital of Victoria because I feel proud of that diversity. And, I think that that's a very healthy way to go in increasingly conservative times. (Interview, 2006)

Michelle's experience is also telling. A heterosexual, 50-something, 30-year resident, farmer and mother, she explained her changed attitude to 'homosexuality' after volunteering at Carnival Day:

> There are lot of people in town who see things in 'black' and 'white'. I think I am one of those, sometimes. I was only at ChillOut as an opportunity to raise funds. Perhaps I am now seeing things in shades of grey. (Interview, 2006)

Her reflections highlight a blurring of the boundaries between straight and gay. Through participating, the foundation of her opposition to non-normative sexualities has become less certain.

For many people from Melbourne, individual negotiations of the flamboyant displays of lesbian and gay sexualities facilitated by ChillOut renders Daylesford–Hepburn Springs 'gay'. Our ChillOut visitors' survey (see Gorman-Murray (2009) for further explanation of method and analysis of the survey data) suggests the festival significantly contributed to the narrative of Daylesford–Hepburn Springs as a 'gay capital' or 'gay heartland' (see also Chapter 1). One question asked of attendees was, 'What does ChillOut for Daylesford–Hepburn Springs's identity?' The largest single group of responses (41%) suggest that ChillOut challenged assumptions of heterosexuality through generating alternative narratives of Daylesford–Hepburn Springs as a gay-friendly place or a gay centre:

> Commonly known as 'gay capital of Victoria'. (Lesbian visitor from Melbourne, 15–25 years)
>
> Promotes Daylesford as being gay friendly. (Lesbian visitor from Melbourne, 26–40 years)
>
> It's the gay capital. (Lesbian visitor from Melbourne, 15–25 years)
>
> Confirms the gay aspect of local identity. (Heterosexual visitor from Melbourne, 26–40 years)
>
> A good reason to visit the gay capital. (Lesbian visitor from Melbourne, 15–25 years)
>
> Solidifies place as a gay icon and gay friendly. (Lesbian visitor from Melbourne, 41–60 years)
>
> It is the gay capital of Victoria country. (Gay visitor from country Victoria, 41–60 years)
>
> The rural gay and lesbian centre of Australia. (Lesbian visitor from Melbourne, 26–40 years)

The popularity of ChillOut amongst lesbian and gay participants from Melbourne sustains imaginaries of Daylesford–Hepburn Springs as the 'gay capital of Victoria'. As a gay capital the town is imbued within discourses of safety, acceptance and tolerance:

> It becomes synonymous with cultural and sexual diversity, acceptance, etc. (Gay visitor from Melbourne, 26–40 years)
>
> Makes it seem tolerant, accepting, good place to come even when ChillOut is not on. (Lesbian visitor from Geelong, 41–60 years)
>
> A gay-friendly town where gays know they are always welcome for a getaway. (Gay visitor from Melbourne, 26–40 years)

What is significant in the above responses is how ChillOut, again, naturalises Daylesford–Hepburn Springs as a particular gay-friendly tourist place beyond the temporal frame of the festival. Further, for those lesbians and gay men living in regional Victoria, ChillOut helps to mobilise narratives of acceptance, help and visibility (see also Gorman-Murray 2009). These discourses became apparent when we asked respondents: 'Should ChillOut be a rural festival?' Responses included:

> Increases access for rural/regional gays. Promotes visible gay presence in regional areas. (Gay visitor from Geelong, 41–60 years)
>
> Good for rural people who may feel isolated. (Lesbian visitor from Geelong, 41–60 years)
>
> Promotes country and its acceptance of gays and lesbians. (Lesbian visitor from country Victoria, 26–40 years)
>
> Helps those more isolated. (Lesbian Hepburn Shire resident, 26–40 years)

These responses highlight how the assumption of heterosexuality still pervades much of regional Australia (cf. Gottschalk & Newton, 2003). For many respondents living in regional Victoria, ChillOut is a red-letter weekend, an event around which a social calendar is planned months, if not a year, in advance.

Given such acknowledgement of the transgressive possibilities of ChillOut, it is perhaps not surprising that the lesbian and gay residents of Daylesford–Hepburn Springs we interviewed avowed that, long after the festival finished, the place remained a 'gay town'. Natasha, a 20-something service worker who grew up in Daylesford, said this is a 'gay town' and 'we feel we belong'. Linda, a 30-something service worker and mother, who moved back to Daylesford 10 years ago, asserted:

Linda: You know it's safe to walk down the street with a same-sex partner holding hands in Daylesford.
Gordon: At any time?
Linda: At any time
Gordon: Not just while the ChillOut festival is running?
Linda: You'd hardly even do that in Oxford Street in Sydney apart from Mardi Gras time, but in Daylesford you could walk down the street, male or female, holding hands with your partner and you probably might get looked at, but you certainly wouldn't get heckled, and the community seem to accept it.
Gordon: And do you think it's in part due to ChillOut?
Linda: Yes, and I would say that's mainly because of the dollar factor for the town. (Interview, 2006)

Linda suggests that in a town reliant upon tourist earnings, the revenue generated from the parties, parades and carnivals of ChillOut help challenge the assumption of everyday space in Daylesford–Hepburn Springs – or at least the main street – is heterosexual.

Unquestionably, ChillOut discursively disrupts the borders that deny the existence of lesbians and gay men in country Australia. Indeed, it is *how* the borders of heterosexuality are disrupted by the ChillOut Festival that has become a source of tension in Daylesford–Hepburn Springs. In a tourist town, ChillOut has located lesbian and gay subjectivities at the intersection of discourses of sexuality, pleasure and economic futures. Hence, Linda also points to the uncertain outcomes of ChillOut: two women holding hands on the main street may remain a disruptive practice, but one that is perhaps sanctioned because of the economic windfall of lesbian and gay tourism and other business activities, a socio-economic frame that positions lesbians and gay men as 'affluent'.

Accepting (particular) Sexual Differences: The Stereotype of the Affluent Gay Tourist

Among those who regarded ChillOut primarily as a tourist attraction, the theme of the 'lesbian tourist' and 'gay tourist' emerged in our empirical materials. Some residents were critical of how ChillOut foregrounded sexuality through economic relationships. While sharing symbols of Pride events – the rainbow flag, in particular – the organising committee does not unite under communal understandings of 'equality', or narratives associated with a history of homophobic discrimination. Amongst residents familiar with the agenda(s) of identity politics and activist coalitions, this ambiguous relationship with Pride is confusing rather than transgressive. For instance, Lucy argued:

> It seems to me that if you raise money, or get grants, no questions are asked, and no principles or guidelines are needed. Why do they [ChillOut] need to make so much money? Drinks at top bar prices, selling water rather than providing it. Buying acceptance? Buying off any local and local verbalisation of homophobia? Why are there no concessions for students, unwaged, family – that is giving back to community or 'charities'? I am surprised that they use the language of 'charity'... . Seems to be some muddled thinking. (Interview, 2006)

Lucy suggested that ChillOut is rendered acceptable to some through its alignment with 'mainstream' neo-liberal economic agendas. Rather than rupturing hegemonic notions of heterosexuality, her critique was that ChillOut expressed sexuality through capitalism rather than political identities. ChillOut, it would appear, is for people who are part of capitalist society. Within the market, sexuality is disciplined by rigid mainstream categories that essentialise all gay men and lesbians as affluent. Lucy's thoughts echo those of Puar (2002) who argued that the market privileges lesbian and gay *consumer* subjectivities. In Halberstam's (2005) terms, ChillOut can be conceptualised as a 'conservative social project' by the way lesbians and gay men are assimilated into the market.

Although we have demonstrated how ChillOut does remake the spatial imaginary of Daylesford–Hepburn Springs as a 'gay capital', at times this acceptance unfurls through market mechanisms. This market-based process often does little to challenge the assumptions of heterosexuality for those lesbians and gay men who are not recognised as 'affluent tourists'. For instance, Phil, a young gay man in his early 20s, exemplified how heterosexism remains resolute in secondary schools, explaining why he left Daylesford because of homophobic assaults:

> I live in a town that is proud, gay-friendly and safe for 4 days in March, and even then only to tourists with money. For the other 361 days, my town, like any other, is fine if you stick with your own kind and conform to predetermined rules in public. I have never been good at these unwritten 'rules'.... Even before I publicly stated my sexuality, I was being harassed for my perceived difference and labelled by the general school population as gay. When I entered secondary education for the first time, it quickly became obvious that my hopes of a more mature, open and accepting environment were unfounded. The experiences of assault within my first two weeks of attendance by a fellow pupil left me crushed, and my trust and belief in humanity and society left tattered and bruised. The immediate response from the school was, of course, to do what many large institutions do when confronted with something of this nature – deny, deny, deny. Soon after this, I left the school, vowing never to return. (Interview, 2006)

Phil identified how, in his experience, ChillOut subverts everyday sexual hierarchies for only a specific group of lesbian and gay people defined through the tourism industry. How his body is sexualised in school by straight classmates remains reliant upon categories defined by heterosexual norms, in an institution not wishing to confront homophobia. His life in Daylesford was made intolerable through homophobic assaults by his peers. Despite the town's image as a 'gay capital', ongoing assumptions of heterosexuality make life just as difficult in Daylesford–Hepburn

Springs as elsewhere, for young people who do not conform to expected gender and sexual norms.

Homophobia is not restricted to young people. The debate that raged around the Council's 2006 'Flag and Banner Protocol' demonstrated that lesbian and gay sexualities remained controversial. The Council had 'unofficially' authorised flags from every festival during the year to fly from the Mayor's balcony of the Town Hall, overlooking Vincent Street – with the exception of a gay-identified rainbow flag, a decision justified by the Council's lack of an official flag protocol. The apparent contradiction – that ChillOut is simultaneously sanctioned and disavowed by Council – is best interpreted by understanding that *where* flags are flown is intrinsically embedded in relationships of power (Leib & Webster, 2004). In Hepburn Shire, it would seem that civic buildings are not 'proper' places to display lesbian and gay symbols. The ChillOut organising committee became increasingly troubled at this situation as homophobia seemed to ultimately underpin the Council's decisions. The protocol introduced after ChillOut in 2006 prohibited the display of any flag or banner other than the Australian national flag from any of the town halls in the Shire. Festival flags could only be flown from a flag pole erected beside the Daylesford Tourist Information Centre. While allies of the ChillOut festival took issue with the flag protocol as failing to represent an inclusive 'community', the decision received support in many letters to the editor in *The Advocate*:

> 'Our town hall represents what our community represents'. What a tragedy that our community now represents the acceptance of the community that accepts that anything goes. The area is known for the gay and lesbian people, for goodness sake. Why do you have to pollute our vision with their flags or can we have some flags for the normal, decent families to let the rest of country know that some decency does still exist? (Treacy, 2006: 8)

This letter highlights a heteronormative assessment of Daylesford–Hepburn Springs: as an 'area known for the gay and lesbian people' who are, in turn, synonymous with pollution, abnormality and moral corruption. Other letters of support for the flag protocol published in *The Advocate* also centred on a refusal to endorse the acceptance of lesbian and gay families as part of the 'normal' community. Letters supporting the refusal to fly the rainbow flag from the Town Hall during ChillOut reinforce straight/gay, local/tourist and everyday/exotic dichotomies. With ChillOut understood by the majority of Councillors as a form of entertainment for gay and lesbian tourists, the ensuing power struggles over where to fly the rainbow flag revealed one way in which the assumption of heterosexuality is maintained in everyday places (Gorman-Murray et al., 2008).

Conclusion

This chapter argues that ChillOut exemplifies the tensions between festivals as 'politics', 'celebration', 'enterprise' and 'fund-raising'. We have demonstrated how for some individuals ChillOut troubles assumptions of heterosexuality that continue to fashion rural Victoria, and yet for others operates to confirm normative assumption of heterosexuality. For those who embrace sexual difference, the pleasure derived from witnessing and performing sexual diversity during ChillOut is central to defying homophobia and heterosexism in everyday life. ChillOut contributes to fashioning Daylesford–Hepburn Springs as a 'gay capital' in heterosexual, lesbian and gay imaginaries. For lesbians and gay men living in regional Victoria, ChillOut provides an important opportunity to generate social networks and perform otherwise masked identities. Yet, even amongst those who embrace sexual diversity there is a sense that ChillOut is only 'buying acceptance' through its deployment within tourism promotion and capitalist relations. Such concerns are exemplified in ongoing experiences of homophobia in schools, the refusal to display the rainbow flag on the town hall and how outside of the festival time-space, publicly accepted expressions of lesbian and gay sexualities along Vincent Street are primarily fashioned through conspicuous consumption.

To conclude, festivals clearly offer ongoing opportunities to lesbian, gay, bisexual, transgender and queer collectives that seek to challenge normative ideas that continue to fashion sexuality beyond metropolitan centres. Festival organisers must remain mindful that these openings arise within the festival time-frame because of how joy facilitates interconnections that may otherwise not be possible. Humour and the deployment of camp, in particular, are essential parts of the entertainment mix, and have a long-standing role in the politics of the gay movement (see Shepherd, 2005). That said, festival organisers must also remain alert to the implications of how celebrations of sexuality become entwined in capitalist relationships, and the repercussions of ticketing costs, re-branding of a festival as a tourist attraction as well as corporate and state sponsorship. Organisers must also remain mindful of such questions as: whose sexualities are being celebrated? Is the festival facilitating rural sexualities, or a metropolitan account of rural sexualities? As festival numbers and costs increase, the event may begin to reflect understandings of sexuality from elsewhere. Equally organisers may wish to consider the implications for teenagers where stigmas are still attached to non-normative sexualities and social support services for young people are not readily accessible.

Note

1. Interviews quoted here were all conducted as part of field work in Hepburn Shire in 2006. Names have been changed to protect anonymity.

References

Alston, M. (1995) *Women on the Land: The Hidden Heart of Rural Australia*. Kensington: UNSW Press.
Anon. (2001) ChillOut Festival promises a weekend of fun for all. *The Advocate* 23 (March), 7.
Anon. (2002) ChillOut brought a warm response. *The Advocate* 13 (March), 28.
Anon. (2006) Chillout in Daylesford. *Bent* 21 (February/March), 67.
Butler, J. (1990) *Gender Trouble*. London: Routledge.
Gervasoni, C. (2005) *Bullboard, Macaronia and Mineral Water: Spa Country's Swiss-Italian Story*. Hepburn Springs: Hepburn Springs Swiss-Italian Festa.
Gorman-Murray, A. (2009) What's the meaning of ChillOut? Rural/urban difference and the cultural significance of Australia's largest rural GLBTQ festival. *Rural Society* 19 (1), 71–86.
Gorman-Murray, A., Waitt, G. and Gibson C. (2008) A queer country? A case study of the politics of gay/lesbian belonging in an Australian country town. *Australian Geographer* 39, 171–191.
Gottschalk, L. and Newton, J. (2003) *Not So Gay in the Bush: 'Coming Out' in Regional and Rural Victoria*. Ballarat: University of Ballarat.
Halberstam, J. (2005) *A Queer Time & Place: Transgender Bodies, Subcultural Lives*. New York: New York University Press.
Johnston, L. (2005) *Queering Tourism. Paradoxical Performances at Gay Pride Parades*. London: Routledge.
Leib, J. and Webster, G. (2004) Banner headlines: The fight over Confederate battle flags in the American South. In D. Janelle, B. Warf and K. Hansen (eds) *WorldMinds: Geographical Perspectives on 100 Problems* (pp. 61–68). Dordrecht: Kluwer.
Markwell, K. (2002) Mardi Gras tourism and the construction of Sydney as an international gay and lesbian city. *GLQ A Journal of Gay and Lesbian Studies* 8, 81–99.
Mulligan, M., James, P., Scanlon, C. and Ziguras, C. (2004) *Creating Resilient Communities: A Comparative Study of 'Sense of Place' and Community Wellbeing in Daylesford and Broadmeadows*. Melbourne: VicHealth.
Munt, S. (1995) The lesbian flaneur. In D. Bell and G. Valentine (eds) *Mapping Desire: Geographies of Sexualities* (pp. 114–125). London: Routledge.
Picard, D. (2006) Gardening the past and being in the world: A popular celebration of the abolition of slavery in La Réunion. In D. Picard and M. Robinson (eds) *Festivals, Tourism and Social Change. Remaking Worlds* (pp. 46–70). Clevedon: Channel View Publications.
Probyn, E. (1993) *Sexing the Self: Gendered Positions in Cultural Studies*. London: Routledge.

Puar, J.K. (2002) Queer tourism: Geographies of globalization. *GLQ, A Journal of Gay and Lesbian Studies* 8, 1–6.
Shepherd, B. (2005) The use of pleasure as a community organizing strategy. *Peace & Change* 30, 435–468.
Treacy, A. (2006) Letter to the editor. *The Advocate* 8 (5 July).
Turner, G. (1994) *Making it National: Nationalism and Australian Popular Culture*. Sydney: Allen and Unwin.
Van Gennep, A. (1960) *The Rites of Passage* (M.B. Vizedome and G.L. Caffee, trans.). London: Routledge (Original publication 1909).

Part 5
Festival People

Chapter 14
Bring in Your Washing: Family Circuses, Festivity and Rural Australia

A. LEMON

While rural festivals are a recent addition to Australia's community development and cultural calendars, family circuses have played a central role in rural communities as a touring 'festival' site for over 150 years, drawing communities together to share a celebratory experience outside the bounds of daily life. Traditional circuses tour cities, country towns and remote settlements, bringing with them a festival atmosphere that might last for a day, a week or longer (St. Leon, 2006). They provide entertainment, distraction from cares and a location for shared camaraderie. Even today, circuses can be the only entertainment to reach remote areas of Australia, drawing audiences to small towns from a radius of some hundreds of kilometres. The peripatetic nature of the circuses, which might visit a town only once every few years, intensifies the festival experience, its presence fostering a bond among people living in remote and rural areas.

When we think of rural festivals, we think of 'place', as these festivals are geographically located, and the majority celebrate a region or town and the community or festival 'audience' that call it home. Circus festivity however inverts geography. The circus does not come *from* a rural community, but *to* it. As Australia's last settler nomadic community, the circus travels, bringing their event to rural and remote towns, creating a 'festival' at each location, making each town feel 'special' for a day or a week. But it is not only the circus performance that creates this festival atmosphere; it is also the circus mythology surrounding it, and the transgressive, nomadic circus culture within which it dwells. Together these three elements – the circus show, mythology and mobile culture – bring an air of excitement, risk and social liberty to rural communities.

Circus History: 'My Family's Been Doin' This Longer Than We Can Remember' (Robert Perry, Alberto's Circus)

Family circuses have been integral to Australian communities since the 1850s (Figure 14.1). During the late 19th and early 20th centuries, they travelled Australia alongside other tented entertainments such as vaudeville, theatre, sideshows and menageries, often featuring at rural agricultural shows. Circuses are unique in being the only travelling entertainment to continue a nomadic role unabated.

Circus history is rich, and deeply entwined with rural Australia. Historian Mark St. Leon (1983: 13) claims that the first known licence to 'perform dancing, tumbling and horsemanship' in NSW was issued in 1830, but by the 1850s numerous circuses were travelling the colonies, performing under canvas, a regular feature in the Antipodean press (Colligan, 1999: 41; St. Leon, 1992). Melbourne's first permanent circus venue, Astley's Amphitheatre, opened in 1854, seating 2000 people (Colligan, 1999: 33). Now known as the Princess Theatre, this venue is still one of Melbourne's major theatres, although no longer used for circus performances.

Circus performance has always acted as a mirror to popular culture, absorbing cultural influences, and reflecting popular cultural fascinations. Early circuses featured equestrian acts, reflecting the central role of horses

Figure 14.1 Lennon Bros Circus, 2007 (Photo: Cal MacKinnon)

in the emerging colonies (St. Leon, 1983). During the gold rushes of the 1850s and 1860s, Japanese and Chinese troupes toured the diggings, bringing with them new acts such as juggling, balancing, top spinning and conjuring (Sissons, 1999: 78). Eleven American circuses toured between 1873 and 1892, introducing travelling menageries, two- and three-ring big tops and grand spectacles (Cannon & St. Leon, 1997: 73; Greaves, 1980: 14; St. Leon, 1983: 76). Australian shows quickly adopted the more acrobatic American equestrian acts, and performed American-influenced 'spectacles' (Brackertz, 1999: 153). And when America's Wild West arrived in 1888, with 'Doc' Carver's team of American cowboys competing in buckjumping events against Australian roughriders (Hicks, 2000: 29; St. Leon, 1992), Australian shows were soon presenting rough riding, whip cracking and sharp shooting, and did so until the mid-20th century. This early American influence on rural Australian entertainment tastes can still be seen with the ongoing popularity of rodeos, campdrafting and country music, all of which are central to many rural festivals (see Chapter 9). Australian circuses began travelling offshore in the 1870s (St. Leon, 1983: 47), touring to Commonwealth countries such as New Zealand, Britain, South Africa, India and parts of South East Asia (St. Leon, 1983: 94).

Circus quickly became big entertainment business. In 1900, Fitzgerald's Circus boasted 118 personnel, 80 horses and ponies and 15 cages of wild animals. In 1901, they erected the 'Olympia' circus building in Melbourne, and in 1902, their show played to a capacity audience of 6000 in Sydney, turning away another 3000 (St. Leon, 1983: 104–109). The outbreak of World War I took its toll, however, with many circuses forced off the road. The Spanish Influenza pandemic followed in 1918–1919, and the government closed entertainment venues prohibiting circuses from travelling, unless in isolated areas. Holden's Circus toured the Aboriginal Pitjantjatjara lands in very remote South Australia for the duration of the pandemic (Holden, 2007), beginning an Australian entertainment tradition soon followed by many artists, including Australian country music legend, Slim Dusty, who began his touring career with circuses, and built his following touring remote Aboriginal towns from the 1950s (Lemon, 1996) (Figure 14.2).

Many elderly circus folk reflect on the years prior to World War II as innocent times, when even small shows were able to 'make an honest living' on the road, appreciated by isolated regional audiences (Ashton, 2007; Bills, 2007; Gill, 2006). This early period also established the intergenerational network of circus families and artists, which still drives Australian circus culture. After the war years, new shows emerged featuring recently arrived European and Eastern European circus artists. Coming to Australia as assisted immigrants, displaced persons, refugees or under

Figure 14.2 Gills Circus/Sideshow line-up board, c. 1940s (Photo: Doyle Gill)

contract to larger shows, they introduced a 'European sensibility' to the often rough and ready style of Australian circus. Economic times were good in the 1950s, and travelling circuses were at their modern zenith as one of Australia's primary family entertainments. By the late 1960s, Australian circus was a hybrid of British, American, European and Asian influences but circus was no longer a carefree life. Television was taking its toll on audience numbers, animal liberation politics had arrived and circuses were feeling the weight of mounting regulations, paperwork and accounting. In the late 1970s, Australian circuses experienced the beginnings of a general downturn that has continued to this day. The introduction of colour television, a nationwide financial downturn and the emergence of 'new circus', notably Circus OZ, attracted a new and younger audience to the circus, but drew them away from the traditional shows (Broadway, 1999). By the 1990s, animal activism had gained momentum, and circuses were a primary target. Activists convinced some local councils to ban circuses; picketed shows; defaced posters and stole animals. By the end of the 20th century, traditional 'exotic' circus animals such as elephants and big cats were a rarity at the circus, and many shows began presenting shows without animals.

In 2009 there were only eight traditional circuses touring Australia full time, and only four – Stardust, Lennon Bros, Perry Bros and Circus Joseph Ashton – traced their circus lineage back to the 19th century. Proprietors of the other permanent shows – Royale, Silver's, Weber Bros and The Great Russian Circus – were of Swiss, German, Spanish and Russian circus descent. Some circuses such as Circus Sunrise, Lorraine Ashton's Classic Circus and Circus Xsavia performed intermittently and a few small shows performed for fairs and festivals. Perry Bros had the only (non-performing) circus elephants. Stardust and Lennon Bros had the only lions and monkeys. Weber Bros and The Great Russian Circus had no animals, and the other shows presented 'domestic animals' such as ponies, dogs, cows, llamas and geese.

Australian circus has absorbed a myriad of cultural influences over the past 170 years, being one of Australia's first 'multicultural' institutions, featuring Aboriginal artists (Figure 14.3) and performers from Asian, African, European, British and American backgrounds. While maintaining their performing traditions, they have adapted to cultural, social and political change. Bureaucracy is the bane of circus life in the 21st century, and animals have become a liability for some. Elephants, once deemed the primary signifier of a successful circus, are no longer performing in Australia. Meanwhile many of the generational circus names are being lost, and with them circus's connection to Australian history.

Circus fortunes mirror those of the communities through which they travel. Their survival is dependent on the financial and emotional well-being of rural and suburban Australia. Circuses still have enormous impact in rural towns, and the more remote the region, the greater the effect. Posters pasted around town raise anticipation. The arrival of animals, caravans and trucks draws children to the lot to watch the circus set-up. The sight of the big top lit up, music drifting on the wind, draws crowds to the show. People come ready to participate in a 170-year Australian tradition that feels brand new each night. After a century and a half of success and struggle, circuses continue to offer festive engagement to rural towns and regions (Figure 14.4).

Figure 14.3 Aboriginal wire dancer, Con Colleano, c. 1930s (Photo: Topsy Hutchens)

Circus Performance: 'We Met Ourselves Coming Through The Back Door' (Dollie Lennon, Lennon Bros Circus)

The traditional circus show is made up of a series of unrelated acts performed in a circus ring, linked by announcements from a ring mistress or ring master. It is often performed by a small number of artists, appearing in numerous acts, frenetically entering and exiting the ring, 'meeting themselves coming through the back door'. An artist's 'star turn' might be on the trapeze, but they will also turn up in the tumbling, the flying, the web and the animal acts.

Traditional circuses are generally family operated, with cast and crew travelling from site to site, living in caravans on the circus site. Family circuses operate as small businesses, and rarely access public funding or other financial support. To research these circuses I undertook oral histories with over 50 elderly circus performers, proprietors and workers across Australia, mainly between 2005 and 2007. I also travelled with three

Figure 14.4 Linda West, Liberty pony act. Lennon Bros Circus, 2007 (Photo: Cal MacKinnon)

circuses. In this time, it became apparent how important circuses are to the local culture in remote areas, and the role they play as a travelling 'festival', as I noted in my research journal:

Fri 12th Aug, 2007 – Mitchell, Queensland.

I make Mitchell by 5 o'clock. I glimpse a gaggle of caravans through the trees, and then the big top looming over them with 'Stardust Circus' picked out in light bulbs along the top. The circus is at the showgrounds, a large swathe of bindies [sharp grass seeds] on the southern edge of town. Show time is 7 o'clock. The canteen and sideshows have already started for the early comers. The sideshows are the usual suspects – laughing clowns, rotating teacups, various other amusements and a van serving fairy floss and dagwood dogs. One circus man is calling out 'Showbaaaags', another 'Every child gets a priiiize'. Everyone in Mitchell is here – kids, mums, dads, teenagers and even a couple so elderly that they creep in on sticks. Inside the big top a wall of animated faces eats fairy floss, and waves glow wands waiting for the show to start. Hundreds of shuffling boots kick up the red dust into a throat-choking cloud. Finally, a fanfare strikes up.

Everyone claps along. The show is about to begin! Everyone's yelling and talking in the dark. An announcement – 'All children please stay in your seats. No children near the ring boxes. Thankyou'. Something dangerous is about to begin!

Circus people find it impossible to find an overarching definition of 'circus'. Like Lorraine Maynard (2007) of Perry Bros, they resort instead to describing their shows as a litany of acts:

> We always had a lot of trick ponies, and mum always did a single trapeze, and dad always did a jockey act, and he always did a trick cycle act. And I learnt the trapeze and to ride the high school horse. I did about four acts. I did the single trapeze, I did double trapeze, I did overland ladder.

Most circus people will draw attention to the range of acts individuals perform, and a 'good' show is described as having pace, entertainment, a focus on the kids, no 'blue' humour and nice costumes. Metaphor in traditional circus shows is an almost unknown concept. Content, form and structure play no part. The show in many ways resembles the culture. As Stoddart remarks:

> Not only has (circus) predominantly sidestepped ... linguistic language for mime, music and physical stunts, it has also traditionally avoided arranging its acts in any kind of narrative form, and has favoured restless, itinerant and temporary structures. (Stoddart, 2000: 5)

Even today, Pepe Ashton describes his Circus Joseph Ashton as 'a variety show, which has one act after the other; performed by our family (Ashton, 2005). You come to our place to see what we can do' he says, 'that's *our* theme' (Westwood, 2004). There are rarely 'fancy' narratives to follow in the traditional circus. The style of the acts might reflect popular cultural interests, but the content will rarely reflect deeply on social issues. It is, as Robert Perry (2005) of Alberto's Circus calls it, 'good, clean, family fun'.

Circus semiotician Peta Tait (2006: 3) suggests circus performance can be read as a metaphor for 'escape from the social order and its regimented identities'. But for the nomadic circus culture, 'escape' from regimented social order is a daily reality. They do not perform their circus acts as a symbol of freedom, but as a daily necessity, a birthright and the engine driving and supporting their survival and identity. And this is what rural audiences come to see – a mythic, nomadic entity performing a dangerous show for their entertainment, enlivening their town for a few days (Figure 14.5).

Figure 14.5 Stardust Circus, Charleville, Queensland, 2005 (Photo: Cal MacKinnon)

Circus Mythology: 'Bring in Your Washing, The Circus is in Town' (Lindsay Lennon, Stardust Circus)

Circus people regularly quote this old saying to underline the attitude circuses frequently face from 'locals', who appear to believe that circus people will steal your washing, and possibly much more. So many tales, tall and sometimes true, are told about the circus it has become a mythic entity. As Stoddart (2000: 2) observes, 'Circus history and circus mythology have become very much entwined', and it can be a challenge to disentangle fact from fiction. The mythic circus is that portrayed in fiction and film, and replicated in the popular media – a world of midgets, bearded ladies, evil ringmasters, maltreated children, cruel animal trainers and criminals running from the law. The mythic circus is tatty and tawdry, glittering and glamorous, flamboyantly transgressive and essentially dangerous. It continually romanticises the circus, or uses it as the setting for evil deeds.

When I attended my first traditional circus show in 2004 – having been a regular audience and creator of contemporary circus for the previous 20 years – I found the traditional circus experience unsettling in its strangeness. Not that the show itself was strange; quite the opposite. The whole event felt too ordinary. I had come to the circus immured in Tait's

(2006: 3) 'phantasmic other' of popular circus mythology. I was seeking a preconceived mythic circus experience born of cinema, books and the popular media. I was unconsciously expecting freak shows and a parade of larger than life characters transgressing social boundaries. But the mythic circus dwells purely in our imaginations, outside time and place, history and geography. It is constructed from historic glimpses, but is not historically accurate. It is an exciting and fearful 'world' that is not geographically fixed. It is hatched in the history of circus lots across the globe, but then nurtured by the cultural imagination. As western culture becomes ever safer and more regulated, the mythic circus becomes more idealised yet denigrated, as we both yearn for and fear lives less constrained by legal institutions and bureaucracies. The mythic circus becomes the place where people can muse upon 'human performances of otherness, and knowing otherness, (can) know themselves' (Broome & Jackomos, 1998: 55).

The press frequently reinforces negative mythic circus stereotypes, particularly the popular misconception that circuses are havens for criminals and vagrants. As the proprietor of Perry Bros Circus, Lorraine Maynard (2007) frequently fields inquiries from local police:

> We had a fellow with us in Townsville, helpin' on the tent. He'd only been with us a week, and the police come down and scruffed him. We didn't know he had a record. It come out in the paper – 'Circus Worker blah blah blah', and it listed all the things he done. And you read it and you thought he'd done all that while he was with us! It's a bit unfair, because if somebody goes out there and rapes a girl, and he works for Woolworths, they won't put in the paper: 'Worker from Woolworths Rapes Girl', will they?!

Numerous circus people recall police visiting as a matter of course when crimes are committed in regions through which they travel. Even so, circus people rarely engage with or publicly deny the negative stereotypes inherent in the mythic circus. They understand that even with its misconceptions and blatant untruths, the mythic circus carves a place for them in broader society's cultural imagination. It gives them the social liberty to continue their transgressive lifestyle. Without the mythology, circus people would be perceived as no more than 'travelling entertainers' – no different from pantomime actors, balloon artists, Returned Services League (RSL) singers and comedians – but circus people are different. They are steeped in circus culture, history and mythology. They grow up keenly aware of the intrigue of their mythology, and even 'factual' stories are 'performances' of circus life; yarns spun to entertain. Yet the reality of circus life in the early 21st century is a far cry from the mythic and media circus. Traditional circuses are strug-

Figure 14.6 Lorraine Ashton's Classic Circus, 2006 (Photo: Andrea Lemon)

gling small businesses concerned with offering good entertainment at cheap prices. Small family groups operate most shows, and care for their animals as extended family members (Figure 14.6). The women are often the business minds, and everyone on the lot works hard to keep the circus juggernaut rolling.

Circus Nomadism: 'A Rolling Stone Gathers No Moss, and a Rolling Home Gathers No Mortgage' (Stafford Bullen, Bullen Bros Circus)

Circus mythology draws rural audiences to the circus site, where the circus ring is the site for public circus performance, but once these audiences are often witness to the private circus world. Circus people do not 'go home' at night. The circus lot is their home, and it is most often on the side of a highway, or on public land, with public facilities. The public walk through the circus lot, share the same toilets, stop and chat, stand and stare. There is often no division in the audiences' mind between public and private. Circus people are often viewed in the same way as circus animals. As clown Gary Grant (2006) observes, 'People think they'll come down here at one o'clock in the mornin' and see a clown still made up havin' a cigarette, or out in the mornin' doin' the washin' with the clown makeup on'. The big top and colourful vans are public signifiers of 'circus'; and any activity on the lot is part of 'the circus', and thus part of the performance. This public performance of private life contributes to the festive atmosphere circuses bring to rural towns, with locals gathering at the lot to watch the unfolding circus life.

Nomadism is a concept fraught with white colonial judgements, and nomadic cultures continue to be viewed by many in the west as more 'primitive' than settled cultures. This colonial 'project of controlling by knowing' (Miller, 1993: 20) continues unabated in the works of contemporary thinkers and writers such as Gilles Deleuze, Felix Guattari and Rosi Braidotti, who appropriate the term nomad to describe liberation from restrictive boundaries placed around schools of thought and theoretical frameworks (Braidotti, 1994; Deleuze & Guattari, 1987). Such authors 'subordinate' nomadic culture to yet another fictive myth in which nomadology is a 'concept (that) must remain pure in order to be useful. (And) in order to remain pure ... has to be "nonactual"' (Miller, 1993: 25). Once

Figure 14.7 Gary Grant, Lorraine Ashton's Classic Circus, 2006 (Photo: Andrea Lemon)

again, nomadic communities and their cultural identity are silenced, this time in favour of a theorised identity (Figure 14.7).

I side with Robyn Davidson (2006: 27) who suggests nomadism is a 'resilient, rational response to ... unpredictable circumstances'. I reclaim this long-derided term as a positive framework within which to understand travelling circus culture, which is influenced by physical geography but not predicated on geographic identity. Circus communities share much with nomadic groups throughout the world. They travel as need necessitates, are familiar with traditional routes and stopping places, give higher priority to blood relations and shared ancestry than economic or social imperatives, usually marry within the extended nomadic community and teach children life skills through participation. They also suffer the mistrust of local communities, restrictions imposed by governments and the bigotry of legal agencies. Circuses depend on family, community and kinship for survival (Gmelch, 1986; Okeley, 1975; Sutherland, 1975). Nomadism has become a cultural footnote in western society, but the circus community has been unique in their ability able to maintain their nomadic culture and identity.

Circus and Place: 'We Didn't Know What To Do with All The Doors' (Robert Perry, Alberto Bros Circus)

Festivals are most often 'placed', celebrating a town, region or nation and the people who 'belong' there. The festive circus experience however does not celebrate place, but brings the community of a place together to celebrate. Circus identity and belonging are not predicated on 'place' – they see themselves first and foremost as 'circus'. Their sense of belonging is built on shared history, culture and identity. It is expressed through the question 'who am I?' not 'where am I?'

The circus is 'country' to its people. It is the 'place' to which they have deep spiritual, historic and cultural attachment. Golda Ashton (2005) recalls literally aching with loneliness when she left Ashton's Circus as a teenager, losing her connection to culture, history, family and identity:

> I missed it straight away to the point where you'd throw up you missed it so much. And how can you replace that big gap in your life? Literally you felt like throwing up because you couldn't fill the laughter. Tele couldn't do it. Goin' to the pub couldn't do it. Makin' friends just wasn't the same, because it's just a laughter that's laughin' at things that happened when I was born. And you can't replace that. It's family laughter. And it's hard to find. And I missed it heaps. And I think it was two years and I'd had enough (Ashton, 2005) (Figure 14.8).

Figure 14.8 Golda Ashton (centre) with nephew, Merrick, and mother, Nikki. Backstage, Circus Joseph Ashton, 2006 (Photo: Andrea Lemon)

All families experience a sense of loss when members leave, but where non-circus families might expect children to leave and pursue marriages or careers, at the circus this can be viewed as betrayal, disrupting the family business, and the core framework of circus culture. After being forced to sell his family's Alberto Bros Circus, Robert Perry (2005) recalls wandering aimlessly around his new house, not knowing what all the doors were for, having lived his entire life on the road until that point:

> My mother was lost when we come here. She was 82. Never lived in a town more'n a week. Well we were lost too. We were all lost together. Lost sheep. It was tough. All the locals talked a different language to us. I'm totally lost with the locals. They talk about flowers (laughs). Talk about cupboards ... concrete. Circus people talk about circus stuff I suppose. Different acts and animals and things that have happened, and people you know. You're right at home. Cos you all know each other.

Circus people do speak a different language, literally and metaphorically from 'locals'. Lorraine Maynard of Perry Bros Circus speaks intriguingly of a constant 'here', which is not where she is when speaking, nor even the place where the circus might be on the day of speaking, but just here – at the circus. Even when speaking of events that took place over 100 years ago, they still took place 'here'. As a researcher, I found myself confused, thinking 'much has happened in Bacchus Marsh over the years'. As a local, my understanding of 'here' is about place, but Lorraine's understanding of 'here' is about circus.

As a nomadic culture, traditional circuses are transgressive by nature. Historian Richard Broome reflects that 'wanderers' are perceived to be beyond moral and social control, and as thus represent 'potential disorder' (Broome & Jackomos, 1998). The circus community continues to be seen as people with 'no fixed address' by most communities through which they pass, as vagrants, criminals and women of loose morals. Bureaucracies legislate against nomadic transgression. The legal system enforces legislation. Circuses are dependent on suburban and rural audiences, bureaucracies and the law to survive.

Audiences might want an 'out of the ordinary' experience when they come to the circus, yet they can also feel threatened by the presence of this nomadic community in their midst. Traditional circuses have frequently been the target of fear-induced violence and discrimination. Circus artists are expected to perform a transgressive, mythic circus in the ring, but once outside the ring they will often underplay their difference, 'passing' as non-transgressive to stay under the bureaucratic, legal and local 'radar'. For 170 years circuses have promoted a public image as ordinary, even conservative people – upholding Christian values, and donating income from shows to local charities (Ashton, 2007). Today, circuses need to be 'out of the ordinary' to draw an audience, but 'regular' enough to survive. Circus artist and proprietor, Pixi Robertson (2005) notes, 'No matter how gorgeous the caravans are, it's still living from day to day ... and on the edge'. Traditional circus life is about daily survival. There is little romance in the transgression of daily nomadic existence. It is hard work. There are fewer conveniences than regulated life, and one must contend with the constant duality of the public's fascination and fear (Figures 14.9 and 14.10).

Circus Survival: 'The Family Needs the Circus, and the Circus Needs the Family' (Stafford Bullen, Bullen Bros Circus)

Circuses currently face numerous challenges to their survival. Some of these challenges are felt across rural Australia, but others are unique to circus culture. Ongoing drought drives up the price of circus animal feed, and affects the rural audiences' ability to pay ticket prices. Rising fuel prices prohibit long haul travel to many isolated areas. Changing attitudes

Figure 14.9 Shannan West with Cassius, Stardust Circus, 2005 (Photo: Cal MacKinnon)

towards performing animals create conflict. Competition for the entertainment dollar contributes to dwindling audience numbers. Spiralling insurance costs and conflicting federal, state and council regulations create an expensive bureaucratic maze out of which some circuses cannot find their way. The nomadic circus lifestyle and cash business is barely tolerated by governments and bureaucracies, and circus people are often viewed with distrust as 'blow ins' by the wider community. Urban, regional and industrial development eats into traditional circus sites, leaving circuses to raise their big tops on marginal lands further from their audience base. According to circus proprietor, Lindsay Lennon (2005), alone over 30% of circus sites in Sydney were lost to urban development between 2004 and 2007. Such challenges are taking their toll on this unique culture, and inhibiting their ability to play their traditional 'festive' role in rural communities.

A surprising number of traditional circuses continue to survive, expressing an ongoing commitment to performing for rural Australia, and

Figure 14.10 Wonona West with Millie, Stardust Circus, 2005 (Photo: Cal MacKinnon)

to continuing their nomadic culture. The family structure of the circus provides a stable base for its members, and the flexibility to survive difficult financial times. And for the family, the circus is meaning. It is the spiritual, mythical and material home which gives its people life. The circus roots its people to family, history, mythology and to 'country'. Circus is a noun, a verb and an adjective. It is where these people live; it is what they do; and it describes their essence. Circus life has never been easy, and life on the road entertaining rural communities continues to be fraught for these nomadic people. While their future may appear to be in jeopardy, their flexibility, unique sense of belonging and extended understanding of community are all key to their survival.

As an art form in which language, gender and class pose no barrier to involvement or enjoyment (Stoddart, 2000), circuses have played a central role in rural Australia since the mid-19th century, drawing communities together to share a celebratory experience outside the bounds of daily life. As travelling spectacles detached from place, circuses bring a festival experience to remote and rural Australia, where audiences gather to witness the circus performance, share in the circus mythology and experience a taste of transgressive nomadic circus culture.

References

Ashton, D. (2007) Interviewed by Andrea Lemon. Circus Xsavia, Royal Melbourne Show, Victoria, 14.9.07.
Ashton, G. and Ashton, S. (2005) Interviewed by Andrea Lemon. Circus Joseph Ashton, Elizabeth, South Australia, 25.11.05.
Ashton, P. and Ashton, M. (2005) Interviewed by Andrea Lemon. Circus Joseph Ashton, Elizabeth, South Australia, 24.11.05.
Bills, D. (2007) Interviewed by Andrea Lemon. Bass, Vic., 14.7.07.
Brackertz, N. (1999) The battle for colonial circus supremacy: John Bull, Uncle Sam and their 'Chariots of Fire'. *Australasian Drama Studies* 35, 145–154.
Braidotti, R. (1994) *Nomadic Subjects: Embodiment and Sexual Difference in Contemporary Feminist Theory*. New York, NY: Columbia University Press.
Broadway, S. (1999) Circus Oz – The first seven years: A memoir. *Australasian Drama Studies* 35, 172–183.
Broome, R. and Jackomos, A. (1998) *Sideshow Alley*. St. Leonards, NSW: Allen & Unwin.
Bullen, S. Interviewed by Judy Cannon. Place and date unknown, Judy Cannon Collection. TRC, National Library of Australia.
Cannon, J. and St. Leon, M. (1997) *Take a Drum and Beat It: The Story of the Astonishing Ashton's*. Sydney: Tytherleigh Press.
Colligan, M. (1999) Circus in theatre: Astley's amphitheatre, Melbourne 1854–1857. *Australasian Drama Studies* 35, 33–43.
Davidson, R. (2006) No fixed address: Nomads and the fate of the planet. *Quarterly Essay* 24 (iii–vi), 1–53.
Deleuze, G. and Guattari, F. (1987) *A Thousand Plateaus*. Minneapolis: University of Minnesota.
Gill, D. (2006) Interviewed by Andrea Lemon. Alberton, Queensland, 16.1.06.
Gmelch, S.B. (1986) Groups that don't want in: Gypsies and other artisan, trader, and entertainer minorities. *Annual Review of Anthropology* 15, 307–330.
Grant, G. (2006) Interviewed by Andrea Lemon. Lorraine Ashton's Classic Circus, Mt. Gravatt, Queensland, 21.1.06.
Greaves, G. (1980) *The Circus Comes to Town: Nostalgia of Australian Big Tops*. Sydney: A.H. and A.W. Reed Pty Ltd.
Hicks, J. (2000) *Australian Cowboys, Roughriders and Rodeos*. Sydney: Angus and Robertson.
Holden, M. and Holden, F. (2007) Interviewed by Andrea Lemon. Victoria, 2007.
Lemon, A. (1996) *Rodeo Girls Go Round the Outside*. Melbourne: McPhee Gribble.
Lennon, D. (2006) Interviewed by Andrea Lemon. Lennon Bros Circus, Chester Hill, NSW, 19.3.06.
Lennon, Jan and Lennon, L. (2005) Interviewed by Andrea Lemon. Stardust Circus, Cunnamulla, Queensland, 17.8.05.
Maynard, L. (2007) Interviewed by Andrea Lemon. Perry Bros Circus, Bacchus Marsh, Vic., 21.2.07.

Miller, C.L. (1993) The postidentitarian predicament in the footnotes of *A Thousand Plateaus*: Nomadology, anthropology, and authority. *Diacritics* 23, 6–35.
Okeley, J. (1975) Gypsies travelling in Southern England. In F. Rehfisch (ed.) *Gypsies, Tinkers and other Travellers* (pp. 55–84). London: Academic Press.
Perry, R. and Perry, B. (2005) Interviewed by Andrea Lemon. Murray Bridge, South Australia, 28.11.05.
Robertson, P. (2005) Interviewed by Andrea Lemon. Melbourne, 8.4.05.
Sissons, D.C.S. (1999) Japanese acrobatic troupes touring Australasia 1867–1900. *Australasian Drama Studies* 35, 73–107.
St. Leon, M. (1983) *Spangles and Sawdust: The Circus in Australia*. Richmond: Greenhouse Publications.
St. Leon, M. (1992) *Index of Australian Show Movements: Principally of Circus and Allied Arts 1833–1956*. St. Leon.
St. Leon, M. (2006) Circus and nation: A critical inquiry into circus in its Australian Setting, 1847–2006, from the perspectives of society, enterprise and culture. School of History and Philosophical Inquiry, University of Sydney, Sydney.
Stoddart, H. (2000) *Rings of Desire: Circus History and Representation*. Manchester: Manchester University Press.
Sutherland, A. (1975) The American Rom: A case of economic adaptation. In F. Rehfisch (ed.) *Gypsies, Tinkers and other Travellers* (pp. 1–40). London: Academic Press.
Tait, P. (2006) Circus bodies defy the risk of falling. Fabulous risk – Online document: On WWW at http://www.semioticon.com/virtuals/circus/index.html
Westwood, M. (2004) Circuses walk the wire. *The Weekend Australian*, 13 March, 16–17.

Chapter 15
Culturing Commitment: Serious Leisure and the Folk Festival Experience

R. BEGG

Introduction

Folk music festivals are remarkably prevalent in Australia, one of the most common and yet poorly understood of all festival types, with a structure and form that differentiates them from rather more commercial festivals. What is different and distinctive about the two folk festivals discussed here is that both were deliberately located away from even small towns, have sought to avoid a commercial ambience and thus, make a more limited contribution to local economy and society than many festivals. That detachment distinguishes folk festivals from the majority, but makes them more like other niche festivals that celebrate particular interests rather than places. While most festivals are seemingly attended by participants on a largely casual basis, for fun and entertainment, with no sense of any particular commitment, for some more specialised festivals many participants may be particularly committed and remain for several days (Mackellar, 2009). This is evident here, and hence, this chapter examines the participation of festival goers in two folk festivals in rural Australia, and particularly focuses on those who demonstrate what can be seen as a high degree of 'commitment' and thus membership of a distinct subculture. Festivals are not merely a few days or hours of frivolity. Specialised forms of participation generally involve high levels of commitment and have been conceptualised through the theory of 'serious leisure' (Stebbins, 1982, 1992, 2001). Focusing on the perceptions and participation of 'serious' festival-goers – those who go regularly – provides insights into the distinct forms of participation and on how festival experiences are generated.

The 'Folk Scene'

The origins of folk music are concerned with ideals of tradition and the 'authentic', even a 'natural style of music played by folk' (Myers, 2004) that kept alive traditions through the performance of spoken word and music. Folk music has been referred to by musicologist, Franco Fabbri (1982: 136) as a social musical genre, which encompasses musical code, rules of behaviour, social relationships, ideological meanings and shared understandings. Categorising folk music in this way indicates how and why a genre that claims a link to authenticity should foster a community of devotees. Such devotees seek to keep alive this 'non-official', independent musical style that is distinguished from commercial genres and professionalism by the ideology of the folk 'community' where music is the product of an ever-evolving communal process. It is regarded as an expression of community interactions, with a porous boundary between performer and audience: 'committedly inclusivist, a celebration of culture from below ... ground[ing] cultural production in community and the face-to-face' (Smith & Brett, 1998: 7). Folk music becomes a musical 'scene' through collective consumption and production of the music. This collective experience can create a heightened social interaction, where participants learn social meanings of the music and generate further meanings themselves. One of the most powerful and effective sites for this to occur is a folk festival.

Australian folk festivals range from small country town festivals to large national ones. Over fifty folk festivals are held annually throughout Australia with new festivals emerging every 'season'. At their core is a focus on 'authenticity'; that is, a liking for 'traditional', especially acoustic, music instead of electronic instruments, and for widespread amateurism instead of 'stars'. Folk festivals in Australia were born out of a strong folk revival in the early 1960s, often established by radical left-wing nationalists, espousing collectivism, egalitarianism and a general suspicion of authority (Smith & Brett, 1998: 4). It was a time when Australians began to look at their roots, encouraging them to find ways to preserve traditions, whether British, European or indeed Australian. Consequently, much folk music in the Australian scene centres on Anglo-Celtic heritage and traditions. Before the advent of the 'festival scene' the culture and traditions of folk music were celebrated through monthly folk club 'come all ye' nights where musical enthusiasts came together to play traditional folk. Increased participation and the subsequent need for larger venues resulted in many of these clubs beginning to organise festivals. By the late 1960s and early 1970s, festivals like the Australian National Folk Festival and the Port Fairy

Folk Festival were becoming increasingly popular and became a space for collective, political and social protest. Today, festivals have become more of a leisure activity than a place of social protest, in some instances a lifestyle pursuit; although, most folk festivals would want to suggest that they nonetheless maintain integrity and stay true to a communal spirit.

This chapter is centred on two small-scale regional festivals in New South Wales at which ethnographic field work was undertaken in 2006, chosen because of a core number of repeat visitors and committed festival-goers at them. Discussion here is based on participant observation, questionnaires and a mixture of unstructured and semi-structured interviews during and after the festivals. The 27th annual St Albans Folk Festival took place in the village of St Albans, about 120 km from Sydney. The festival drew performers from all over Australia, but with a large number of regionally based acts. It has come a long way from its early beginnings in 1979, when a group of locals began playing music in the village's only pub, The Settler's Arms. The festival organisers have resisted many pushes from 'outsiders' to turn it into a major commercial event, allowing its traditions and reputation to continue unchanged. The festival is small in terms of visitor numbers, in comparison to others on the festival circuit, averaging around 600 campers and 400 hundred day-trippers over the five days, which constituted a record attendance. The festival layout contributes towards a relaxed village atmosphere, with performances taking place on a main stage, in a local barn, a cafe, the church and the Settler's Arms Pub. By contrast, the Kangaroo Valley Folk Festival was new, operated by the Shoalhaven Folk Club, and located in and around the Kangaroo Valley showground, approximately 200 km from Sydney. The festival attracted about 600 people and a large number of South Coast and national performers, with a programme that was common to most folk festivals and included musicians, buskers, dancers, poets, workshops, folk dances, blackboard venues and special school concerts. The layout of the festival consisted of two large performance marquees, and made use of the church, community hall, The Friendly Inn Pub and a local cafe for the poet's breakfasts.

Both festivals attracted a range of participants, many of whom stayed on the sites, which consequently took on the characteristics of semi-self-contained villages, composed of tents and camper vans, which further separated committed festival-goers from the world beyond. At St Albans and Kangaroo Valley respectively, 76% and 80% identified themselves as general festival-goers, many of whom stayed for no more than a day of the five-day and three-day long festivals. At both festivals more than 50% of participants were over 50, an older group than at most festivals; most were in full-time employment and more than 63% had received tertiary education, again a very high proportion. Most travelled with family and friends from nearby parts of New South Wales and saw the festivals as an opportunity for quality family time. Most festival-goers were drawn by the joint attraction of 'music and social interaction' with many observing 'you can't have one without the other'. Folk music demanded interaction rather than passive listening. The other key influence was the desire to escape daily routines. Day-trippers were more likely to emphasise pleasure and entertainment. The same structure also characterises the Palm Creek Folk Festival in Queensland (Begg, 2006). Most festival-goers, even at folk festivals, are casual visitors. Many day-trippers had attended no other festivals in the previous year, but some more committed festival-goers had been to as many as 28. Many of these latter followed a regular folk festival circuit. Some 38% of participants had attended the St Albans Festival more than six times, and the duration of expressed involvement in the folk scene ranged from none to 52 years. At Kangaroo valley involvement ranged from one to 43 years. As an established festival St Albans attracted the more committed festival-goers, many of whom stayed throughout the five days.

The remainder of this chapter focuses on the more committed festival-goers, the serious participants who usually stayed for the entirety of the festival, were regular attendees, both at St. Albans and at other festivals on the circuit, and for whom participating in the folk scene was a way of life. While such serious festival-goers are only a small proportion of the hundreds of participants, they are at the core of the festival and are central to its ethos. A flavour of this is presented in Box 15.1.

Box 15.1 A weekend in the life of a folkie

A couple arrives while the festival is still being set up, and immediately take care to park by the communal fire spot at the night owls campsite, if their festival friends have not already saved them a spot. He is dressed in a pair of worn out jeans and is likely to sport a ragged akubra hat that has travelled to every festival. His guitar is safely protected in a battered guitar case covered with festival stickers from all over Australia. She may wear a dress and a pair of her favourite 'festival sandals'. The routine 'tent set up' happens almost effortlessly, and is followed by the construction of their 'festival living-room' made from an oversized tarp that is tied to the side of the van and the tree branches above. Still suffering from a 'hangover of their daily lives' they construct a clear physical boundary around their personal campsite, temporarily forgetting they are 'in a place where none of this stuff matters'.

There is really no need for them to buy a program since they already know everybody playing. There is no need for him to buy a folk festival T-shirt, since he is already wearing a faded one from the same festival held five years ago. The older and more distant T-shirts are highly prized status symbols. Around the festival grounds, they are in their element, warmly embracing old friends who also travel the folk festival circuit. If it is early in the season, they may not have seen each other for nearly a year, and there is a lot of catching up to do. While cruising the festival grounds, they may pass various food and craft stalls, but he naturally gravitates to the instrument maker, with whom he chats knowledgeably. Generally avoiding the local Scouts sausage sandwiches, she returns to the campsite for a pre-made meal around noon. Indeed, while relaxing, their 'living-room' may become the site for an impromptu jam session.

The entire day has the character of a family reunion, of a gathering of the clan. It is an extended family devoted to folk music as a social symbol for a particular lifestyle. The music at the evening concert is expected in some degree to carry, articulate and express this entire lifestyle. But oddly enough, the folkies may not really pay much attention to the music itself at the concert. They already know the music. They have significant collections of records, tapes and CDs at home. The evening concert, like the rest of the day, is another opportunity to socialise with new and old friends. Collectively, it becomes a stage for the folkies to act out their lifestyle and values. However, the main event of the day is the informal musical jam session held late that night.

These customary events have the appearance of spontaneity, but are actually highly structured. Participation in them demonstrates the folkies' skills and knowledge of folk music and traditions. The rules for participation are never made explicit, but they are still present. It is best to bring along a six-pack of beer or a 'good drop of homebrew'. Bringing along a generous supply of an acceptable beverage is a good entrée into a small group. The newcomer must not be too eager to join in but must hang back until the appropriate moment, often cued by a nod from someone already playing. Even then, the newcomer must not play too loud or too long, or the group will just walk away.

As the night slips away, songs are exchanged, poems are told, the fire is kept alight, and the beer is consumed. The later one stays up,

> the better. It is not unusual to see 'jammers' straggling back to their tents and sleeping bags with the first light of dawn. Indeed the most prestige accrues to those who do not go to bed at all. Predictably, the folkie is exhausted by the end of a three-day festival. Before he leaves the festival grounds, he picks up a few T-shirts. By now they are being sold at half price. She also remembers to pick up a discarded program to tuck into his guitar case to save as a souvenir. The trip back to the 'everyday' is spent reliving their festival experiences and planning for the next festival.

Folk 'Festivaling' and Subcultural Commitment

Serious festival participation occurs in a space where social divisions (based on role, status, class, gender, ethnicity, age) are, to varying degrees, suspended. Homogeneity develops between those sharing the experiences of cooking, camping, eating, playing and singing, workshopping: in effect folk festivaling. Festivaling demonstrates how individual pursuits combine to form collective activities, which in turn strengthen social relations, while demonstrations of commitment result in the formation and continuation of social groupings. Performing, committing time and staying overnight are the primary indicators of serious commitment, and all serious festival goers refer to themselves as 'folkies'. David, one self-identified folkie, commented 'there are two types of people that attend folk festivals: the folkies and the public'.

Regular festival participation coincides with the concept of serious leisure, defined as 'the systematic pursuit of an amateur, hobbyist, or volunteer activity that is sufficiently substantial and interesting for the participant to find a career there in acquisition and expression of its special skills and knowledge' (Stebbins, 1992: 3). 'Seriousness' is demonstrated by repeat and regular participation, definite involvement rather than casual participation, and an enduring interest in folk music and festival life, all of which demonstrate commitment. Commitment is also measured by the significant investments of time, money, energy and emotion that participants place on their festival involvement. Commitment is also expressed through a 'folk festival career' reflecting the acquisition and expression of particular skills and knowledge associated with various aspects of folk music, performance and its accompanying traditions. This commitment and significant personal effort was reiterated by Robin, a folkie of 19 years who suggested, 'the people who dominate in this folk scene are those who put in the time and effort to perfect their instruments and performance, learn tunes and have the best knowledge of all the songs'. Ian, a regular at St Albans, noted, 'folkies are the definition of commitment', while Campbell, regularly known as 'one of the longest serving folkies on the circuit' said, 'these festivals are my home, they're what I do and pretty much the reason I exist'. This kind of festival participation exemplifies Stebbins' (1996) conceptualisation of serious tourists as 'those for whom cultural pursuits are an active form of identity creation, an extension of general leisure and a systematic pursuit'. Festival-goers are much the same as serious tourists.

A 'festival career' evolves through a number of achievements, particularly those associated with elements of performance (such as regular involvement in jam sessions) and social group identification, which create a range of personal benefits for participants. The most prominent benefit is a sense of belonging, through regular social interaction based on shared interests, common feelings and general like-mindedness. Jack, a 'well-travelled folkie of 32 years' attending the Kangaroo Valley festival regarded 'festival participation as a bonding opportunity for like-minded people ... it's like a family reunion'. James, a regular at St. Albans echoed this by suggesting regular folk festival attendance provided a sense of belonging to 'a real family style community dedicated to the natural sounds of folk'. This sense of belonging also results in the extension of many social networks outside the festival space, through monthly gatherings at folk clubs, pub jam nights and various forms of communication for geographically dispersed groups. Griz, 'a folkie first and home-brew specialist second' said that 'most of us are members of folk clubs or go to a lot of the local pub sessions. So we always see other jammers around the traps when there are no festivals to go to'. Other personal benefits commonly associated with these festival experiences included: self renewal resulting from feelings of relaxation within the festival space; self expression through music and performance, which in turn had positive affects on self-esteem; and finally, subcultural capital (i.e. status symbols) reflected through the attainment of lasting physical products (e.g. collections of folk memorabilia, including rare types of instruments and festival T-shirts). Steve, a regular on the folk circuit summed up his festival experiences by suggesting that 'it's all about meeting people and stimulating my mind ... every festival is a learning experience for me, whether it be about the music, learning a new set of chords or simply learning more about me and life ... these things open up your horizons'.

Collective feelings of belonging and solidarity among serious festival-goers result in the growth of a unique ethos centred around developed 'social worlds' (Unruh, 1980) and corresponding identities. At an individual level, identification as a 'folkie' denotes a devotion to folk music characterised by an enduring commitment to folk festivals. Collectively, 'folkies' represent a folk community that is composed of special beliefs, traditions, values, norms and performance standards. The formation of the folk community parallels the evolution of a 'strong subculture' that Stebbins (1992, 1993) suggests is associated with serious leisure activities. It is these serious leisure activities then, which tend to serve as a catalyst to the development of a unique social organisation unlike casual leisure. Serious participation provides 'folkies' with an opportunity to acquire and express a wide range of skills and knowledge that are particular to various facets of the folk community, whether music, dance or spoken word. These sometimes subtle distinctions divide the folk community into a number of identifiable social groups according to a shared commitment towards a particular activity or interest. Festival 'families' were one such group that generally encompassed strong personal connections and the formation of lasting friendships. Most serious participants belonged to particular folk families that 'adopted' and acquired them during the course of repeated festival visits.

Subcultural Style

Within the wider context of being a folkie and within the festivals other smaller groups – subcultures – could be identified, which evolved through the formation and performance of specific and sometimes distinct identities. Seven such social groups were present at the two festivals: Beardies; Jammers; Irish Fiddlers; Poets; Dancers; Campers and Vanners. 'Serious' festival-goers identified themselves with particular groups, based on a mixture of shared interests and values. These groups are not mutually exclusive but rather encourage cross-membership, bringing about a polycentredness in which individuals can belong to multiple groups. The loose nature of the groups enables serious festival-goers to pursue a number of interests during the festival, and through their involvement accrue subcultural capital (i.e. status, skills and experience) during their 'festival career'. But above all it is the committed nature of these festival participants that both constructs the atmosphere of folk festivals and helps to sustain the traditional elements of folk music and its community.

Sub-cultures have distinct styles and compositions. The Beardies are men, mainly heavily bearded; described by David as 'the traditionalists and fundamentalists of the folk scene' who are often heads of folk club, the older generation and the highly respected (Figure 15.1). They dominate jam sessions by choosing traditional interpretations and favour

Figure 15.1 'Committed folkie', St Albans Folk Festival 2006 (Photo: Robbie Begg)

acoustic styles. Jammers are active performers as singers or musicians in jam sessions, and have usually moved from being involved in 'open' sessions, where all are welcome, to 'closed' sessions where serious participants play together and a particular performance etiquette is expected (Figure 15.2). Irish Fiddlers constitute a smaller group with usually Celtic heritage who tend to be more exclusive than other groups and hold distinctive sessions fuelled by Guinness that establish a Celtic presence at the festivals (Figure 15.3). Poets usually gather in the mornings for the poets' breakfast, where poetry is performed, much with romanticised notions of the Australian bush, settlers and farmers. Poets swap old and new poems and the final morning of the festival is usually reserved for the poets' breakfast on the main stage. Dancers were usually groups that had formed and practised outside the festivals and then performed at the festival. Three dance forms were prominent – Morris dance (which often involved British expatriates), Bush Dance and mediaeval ritual dance – some of

Culturing Commitment 257

Figure 15.2 'Night owl' jam session, St Albans Folk Festival 2006 (Photo: Robbie Begg)

which were also performed at informal sessions (Figure 15.4). The campgrounds at festivals effectively become divided into social groups where Campers and Vanners are distinctive groups. The Vanners were seen as a group who spent less time with others because they had brought their own social space with them, and their vans were often too large to enter particular parts of the site. Campers occupy 'night owl' and 'early bird' sites according to whether they come with families or wish to be involved in late night sessions. Such groups are thus differentiated in various ways according to the nature of their participation, their experience and even their on-site residential location, and thus they overlap significantly enabling serious participants to alternate between groups and be temporary members of several groups.

Over time, commitment may become greater and change form. John, a folkie for 32 years, said 'once your guitar playing skills get good and you're confident in your knowledge and ability, you feel like you're "in" ... and so you're able to relax a bit more and take more interest in getting to know people better rather than just coming out to practice and learn new songs'. Serious participants become self-described 'folk nomads' on the festival

Figure 15.3 Blackboard Session at The Friendly Inn Pub, Kangaroo Valley Folk Festival 2006 (Photo: Robbie Begg)

circuit. Festivals were learning experiences. Serious participants become more involved in social networks, jam sessions, and the wider camaraderie of festivals. Some may persevere through adversity – domestic conflicts over time 'wasted' or not spent together, costs, unmet work commitments – but 'we wouldn't have it any other way because that's the price you pay for such wonderful memories and experiences'. Indeed the occasional dissent from the everyday world may harden the resolve to commitment. Folk families even began to replace 'real' families in terms of the strength of social ties and support, as Jane put it, in 'a real family style community dedicated to the natural sounds of folk'. Groups could become emotional communities, centred on commitment, equality, acceptance and enjoyment. Robin found that social interactions at the festivals 'allow me to find myself and rekindle who I am ... they give me hope that good people still exist in such a selfish world'. Perseverance and the gradual acquisition of skills and knowledge are ultimately highly rewarding.

The enduring nature of folk festival participation and its qualities of serious leisure exemplify a 'culture of commitment' (Tomlinson, 1993). At

Figure 15.4 Mediaeval dancers, St Albans Folk Festival 2006 (Photo: Robbie Begg)

one level, this culture is devoted to the continuation of 'traditional' folk music and its associated lifestyle. At another, it represents an active pursuit of experiences that provides personal benefits of self-renewal, self-expression, self-enrichment and social interaction, generally lacking in everyday life. These cultures of commitment can be seen as particular responses to new social and cultural situations and circumstances, which can increasingly serve as substitute lifestyles, identities and central life interests (Stebbins, 2001) within a sometimes threatening, unknown and impersonal world. Serious leisure stresses individual commitment to a chosen pursuit, organised along collective groupings, and offering both intimate and communal cultural experiences. Festival participation as a form of serious leisure offers an enhanced sense of personal status, cultural identity and group membership.

Serious participation in festivals and all leisure activities results in distinctive social worlds, social networks, specific lifestyles and multiple small groups, around a central collective activity. Subcultures involve groups of people that have something in common with each other and which distinguish them in some way from others. Hebdige (1979), who pioneered the analysis of subcultural style, used the example of punks, describing them as bricoleurs, young people who used a heterogeneous mixture of things to create a collage that became the basis of a distinct and 'homologous style'. Willis (1978: 63), who first applied the term 'homology' to subculture in his study of 1970s hippies and bikers, used it to describe 'the symbolic fit between the values and lifestyles of a group, its subjective experience and the musical forms it uses to express or reinforce its focal concerns'. The bricolage 'effect' and the formation of homologous meanings create and constitute subcultural identities.

Folk festivaling resembles the bricolage of a variety of interests, practices and activities that form the 'folkie' subcultural identity. At one level, regular festival participation is motivated by a collective interest in folk music, but the practice of folk music is divided into a range of specific individual interests. The majority of participants use their festival attendance as a means of acquiring and expressing skills and knowledge regarding their folk interests. Interests range from specific instruments, to particular types of dance or spoken word. At another level, this form of participation involves specific interests within the festival lifestyle (i.e. camping and social interaction). Together, these interests form collective activities, representing expressions of commitment, which are organised around immediate social group identification.

The 'ordering' of these practices and activities into a homology represents expressions of values and norms, which serve to reinforce the folk community's focal concerns. Folk performances and the accompanied social interactions, serve as an active role for the preservation and continuation of oral traditions. They also provide participants with a sense of belonging through family togetherness and a shared festival community. Moreover, the 'liminality of festivals, being outside or on the peripheries of everyday life' (Turner, 1974: 47), the impromptu nature of particular folk performance, and the simplicity of festival life encompass much celebrated values of personal freedom. 'Festivaling' therefore involves both individual and collective acts of commitment, which become expressions of a subcultural style.

Historically, traditions of folk music advocate ideals of collectivism, egalitarianism and a general suspicion of dominant systems, particularly evident through the meanings of folk song. Traditional subcultural themes of alienation, resistance and protest also underscore much of folk music's history. In Australia, folk festivals became the sites for collective social protests; many contemporary participants at the festivals traced personal connections to the anti-Vietnam protests of the late 1960s and early 1970s. Collectively, these acts and the associated forms of participation came to resemble a 'deviant' subcultural style. While a serious involvement within the festival scene held connections to various alienating experiences in everyday life, a need to temporarily escape routine structures (i.e. home and work life) and, more importantly, the pursuit of alternative experiences and the lifestyle remained most relevant. Although certain elements of subcultural theory remain applicable to festivaling, the shifting nature of folk culture as an alternative lifestyle rather than a medium for social protest and acts of resistance, suggests that traditional notions of subculture, centred on class and workplace, are less relevant and indeed dated.

While social divisions like class, gender and ethnicity may play a lesser role in contemporary cultures, they remain of importance. For instance, despite serious festival-goers' shared interests in folk music and festival life, regular participation requires large amounts of travel, time off work and a number of associated costs, reflecting particular class positions where such recurring costs can be supported. Furthermore, identities are 'contingent on sets of structural and material factors' and are 'embedded in the social fabric' of 'real social worlds' (Ball *et al.*, 2000: 57). Influences from ethnicity also have sustained relevance, with much Australian folk music shaped by its Anglo-Celtic origins, and many of the serious participants regarding festivals as celebrations of ethnicity. Stan, a regular at St Albans, suggested, 'the concentrated Anglo-Celtic "flavour" present at most of these festivals is maintained through our strong values of preserving tradition and as a result the festivals become a celebration of our ethnicity'. So, while contemporary leisure cultures appear to be moving away from the fixity of social divisions, these factors still remain as underlying signifiers of group identities. Moreover, while distinctive elements of class and ethnicity still underlie serious participation, the 'temporariness' of festival life, and the loosely bound characteristics of immediate social groups suggest a relatively 'fluid' subcultural formation, but at the same time one that extends 'backwards' into everyday life.

Viewing such subcultural groups through a lens of *neo-tribalism* (Maffesoli, 1996) suggests a series of temporal gatherings characterised by fluid boundaries and floating memberships. An individual's 'folk' identity represents only one temporary social network, thus the folk community reflects a transitory neo-tribal group, which is neither fixed nor permanent, but involves a constant back and forth between 'tribe' and masses (Bennett, 1999). The flow between multiple signs of identity is evident within the 'temporary' festival space between the serious participants' immediate social groupings (i.e. jammers, poets and dancers), where participants can have a number of belongings, and in contrast to the more casual context of the wider festival.

The folk scene is characterised by a set of values, particular styles and learnt skills represented through music, dance and the spoken word. These distinctive characteristics are interconnected to some extent and differentiate serious festival-goers from casual visitors during festivals. A festival-goer is distinguishable from a casual visitor through elements of their 'career' characterised by acts of perseverance (i.e. regularly attending festivals), significant personal effort (i.e. the attainment of skills and knowledge) and the subcultural capital (i.e. festival status) derived from a number of durable benefits. These distinguishing characteristics rely on a degree of commitment, partly enhanced through the majority of festivals on the Australian folk circuit being organised and run by folk clubs and the folk community. This tends to limit the extent of commercialisation and provide a space for continuing folk traditions.

The Final Chord

Folk festivals are distinctive in their at least partial isolation from the towns and communities that so often sponsor and support festivals, including many other music festivals. That isolation places the folk festivals in liminal spaces and plays a significant part in enabling and encouraging many participants to make a serious commitment to involvement, without distraction by everyday life. That commitment extended beyond the immediate festival, through involvement in folk clubs and attending a series of festivals as 'nomads' on the 'folk circuit'. Such serious participants identified themselves as not merely casual festival-goers, but first and foremost as folkies and secondarily as members of particular subgroups, albeit flexible and overlapping. Identification as 'folkies' and involvement with subcultural groups provides such participants with an intense sense of belonging, particular personal benefits and even a rationale for life, especially where that commitment extended back to the Vietnam war protests of the 1970s. Commitment is enhanced by residence on site and by the relatively small size of the folk festivals, where people are amongst friends and 'family'.

Demonstrations of commitment emerge through a progressive 'festival career' involving the attainment and expression of skills and knowledge particular to folk music, forms of performance and the associated traditions. A devotion to festival life is also motivated by the sense of enjoyment, fun and relaxation. As everywhere else, folk festivals too are about enjoyment. The extent to which serious participants attain authentic self and interpersonal relationships positively reflects their level of involvement while social relations among serious participants stimulate an 'emotional community'. Social relationships and the 'liminal' nature of the festival space, being 'betwixt and between', intensify feelings of inclusion, belonging and being valued. Serious participation represents more of a pursuit of an alternative lifestyle, rather than an escape from alienating experiences of everyday life.

Folk communities are not rooted in one place but remain consistent from festival to festival. Perhaps certain forms of music that enhance 'traditional' values even 'call out' to geographically dispersed audiences, bringing them together to form collective groupings (Slobin, 1993). Communal practices of music and social interactions reflect an active role in preserving the heritage and traditional roots of folk music and its culture. In spite of being geographically dispersed, folkies share common interests, hold feelings of commitment and identity towards one another, that translate into both a translocal sense of identity and a consistent set of values and styles. At the core of folk festivals are a large number of enthusiasts, who actively shape and construct the festival, give it direction and vitality and for whom it is a significant part of their lives. Such enthusiasts may very well be at the core of other niche festivals.

References

Ball, S., Maguire, M. and Macrae, S. (2000) *Choice, Pathways and Transitions Post-16: New Youth, New Economies in the Global City*. London: Routledge.
Begg, R. (2006) Culturing commitment: Serious leisure and the folk festival experience. Honours Thesis, University of Sydney.
Bennett, A. (1999) Subcultures or neo-tribes?: Rethinking the relationship between youth, style and musical taste. *Sociology* 33, 255–259.
Fabbri, F. (1982) What kind of music. *Popular Music* 2, 131–144.
Hebdige, D. (1979) *Subculture: The Meaning of Style*. London: Routledge.
Mackellar, J. (2009) Dabblers, fans and fanatics: Exploring behavioural segmentation at a special interest event. *Journal of Vacation Marketing* 15, 5–24.
Maffesoli, M. (1996) *The Times of the Tribes: The Decline of Individualism in Mass Society*. London: Sage.
Myers, D. (2004) *A Score and a Half of Folk: Thirty Years of the Monaro Folk Music Society Inc.* Pearce, ACT: Sefton Publications.
Slobin, M. (1993) *Subcultural Sounds: Micromusics of the West*. Hanover: Wesleyan University Press.
Smith, G. and Brett, J. (1998) Nation, authenticity and social difference in Australian popular music: Folk, country, multicultural. *Journal of Australian Studies* 56, 3–17.
Stebbins, R. (1982) Serious leisure: A conceptual statement. *Pacific Sociological Review* 25, 251–272.

Stebbins, R. (1992) *Amateurs, Professionals, and Serious Leisure*. Montreal: McGill Queen's University Press.
Stebbins, R. (1993) Social world, lifestyle and serious leisure: Toward a mesostructural analysis. *World Leisure and Recreation* 35, 23–26.
Stebbins, R. (1996) Cultural tourism as serious leisure. *Annals of Tourism Research* 24, 450–452.
Stebbins, R. (2001) *New Directions in the Theory and Research of Serious Leisure*. New York, NY: The Edwin Mellen Press.
Tomlinson, A. (1993) Culture of commitment in leisure: Notes towards the understanding of a serious legacy. *World Leisure and Recreation Association* 35, 6–13.
Turner, V. (1974) Liminal to liminoid in play, flow, and ritual: An essay in comparative symbology. *Rice University Studies* 60, 53–92.
Unruh, D.R. (1980) The nature of social worlds. *Pacific Sociological Review* 23, 271–296.
Willis, P. (1978) *Profane Culture*. London: Routledge.

Chapter 16
Tartans, Kilts and Bagpipes: Cultural Identity and Community Creation at the Bundanoon is Brigadoon Scottish Festival

B. RUTING and J. LI

According to Scottish folklore, Brigadoon is a mystical village in the Scottish Highlands that was locked in magical spell long ago. It appears out of the mist for one day every century. It is a mystical place from a bygone era, filled with the sounds of bagpipes and the sights of kilts and bonnie lasses. At the end of the day, it disappears back into the mist.

The Scottish Highlands are not the only place where old traditions and a special way of life seem to have been lost. In recent decades, rural Australia has also undergone drastic social and economic changes. It is no longer the iconic place of Australian folklore. Agricultural production has changed, jobs have been lost and populations have declined in many towns. Some places have attracted 'seachangers' and 'treechangers,' and others have turned to niche food production or tourism. Yet as places across the globe are transformed for the tourist market, many smaller towns find it hard to compete and are left behind.

Festivals sometimes offer a solution. The small town of Bundanoon in the NSW Southern Highlands offers one example. Although it has become popular with 'treechangers' and has long been a place of rural relaxation for stressed city folk, it has also emerged as the home of one of Australia's largest ethnic festivals. Bundanoon hosts *Brigadoon*, an annual one-day Scottish festival. This festival has partly shaped senses of identity and community, not just for Bundanoon's residents, but also for the thousands of visitors that attend the festival each year.

In this chapter, we examine the distinctive and successful Bundanoon is Brigadoon Annual Highland Gathering, reputedly the largest Scottish festival in the Southern Hemisphere. It attracts over 10,000 visitors

Figure 16.1 Caber tossing, Bundanoon is Brigadoon Scottish Festival, 2006 (Photo: Brad Ruting)

annually, mostly from Sydney and Canberra. They come to watch or participate in distinctly Scottish performances and activities, including bagpipe bands, a parade of clan societies, dancing troupes and a range of sports, from shot-putt and boulder lifting to caber tossing and haggis hurling. Tartans, kilts, haggis and shortbread abound.

We use the themes of identity and community to investigate why this festival – which we visited in 2006, 2007 and 2008 – is so popular and successful. We examine the festival's relationship to its host town and how it has been used to both engage local community groups and establish a distinctive place identity. We then explore the range of motivations people have for attending, from the superficial to the deeply emotive. This festival has special meanings for many people who consider themselves part of a broad Scottish 'community' in Australia (Figure 16.1).

Local Impacts and Place Identity

An estimated 35 million people of Scottish ancestry live outside Scotland, more than seven times the population of Scotland itself. Scottish festivals and Highland games are prevalent throughout England, the United States, Canada, Australia and other places with histories of Scottish migration. Some Scottish festivals in North America have attracted researchers' attention (Chhabra *et al.*, 2003; Ray, 2005), but as yet there has been no research on such festivals in Australia.

One of the strangest aspects of *Brigadoon* is that it is held in Bundanoon. Bundanoon is a town of approximately 2000 people, 128 km south-west of Sydney in the Southern Highlands region of NSW. Although only one-hour's drive from Canberra, and about two hours from Sydney, Bundanoon is a small rural town with no significant history of Scottish immigration or settlement. The town has several tourist attractions, including a national park and health spas, but its main drawcard is the festival.

Brigadoon has been held every April since 1978. Attendance was around 10,000 in 2009, although it has averaged around 14,000 for the past decade. Even in 2007, 12,000 visitors attended despite heavy rain and much mud on the oval (many felt that this was authentically Scottish weather!). The festival is organised and run by volunteers from the local region, with the theme based on the 1947 musical *Brigadoon* (itself based on older folklore). The misty village of fantasy and romance that featured in the musical is recreated at Bundanoon, with fog occasionally shrouding the street parade which is held at the start of each festival (Figure 16.2). The parade features marching bagpipe bands, clan societies and floats from local community organisations and schools. It ends at the local oval, where the festival is held. Around one in 10 visitors wear a kilt, and over half wear tartan in some form.

As festivals go, the event is managed well and has many benefits for its host region. It is controlled by a board of volunteer members, most of whom live locally. It is non-profit, with after-cost revenues divided up among the local community groups that volunteer labour to set up and run the festival (e.g. $47,400 was distributed to volunteer groups after the 2009 festival, plus a further $12,700 in donations to participating bands). Around 300 volunteers have been involved each year in recent festivals, donating around 2000–3000 labour hours by preparing the oval, marshalling visitors, setting up stalls and providing other services. Members of the organising committee devote substantial amounts of their own time, with some working full-time for several months before each festival. Stallholder fees, corporate sponsorships and entry fees ($18 for adults, $5 for children) are the main sources of income. Public liability insurance and tent hire are the biggest costs.

Brigadoon is also supported by other organisations. The Fire Brigade and State Emergency Service block off several streets while the festival is on, and the local council allows the use of the sports oval and provides car

Figure 16.2 The parade down Erith Street, Bundanoon, 2007 (Photo: Brad Ruting)

spaces and waste collection. As *Brigadoon* is mostly confined to the oval, its environmental impacts are minimal.

Although only a one-day event, *Brigadoon* has noticeable direct and indirect impacts on Bundanoon. It is the town's largest tourist attraction, which has given Bundanoon a reputation further afield as a 'Scottish' place. Like many festivals, the economic benefits of *Brigadoon* are highly concentrated, in both temporal and spatial terms. Stallholders at the festival report doing good business. The festival is the biggest fundraising event of the year for local groups such as schools, churches and the Rotary Club. However, many stallholders come in from out of town and travel to Bundanoon only for the day of the festival. Most visitors do not stay overnight in Bundanoon itself (due to the limited accommodation available and Bundanoon's proximity to Sydney and Canberra) and few spend much time on the town's main street.

Some businesses in Bundanoon do well, but since the main street is around 500 m walk from the festival grounds (across the railway line) and the festival is largely self-contained (with entertainment, food, drinks and toilet facilities available), only a few cafés and the newsagency report above-average turnover. Some cafés put on extra staff on the festival day but most other businesses do not. A handful of shops reported that some local customers stay away during the festival, due to a lack of parking and the crowded main street. Some others reported falling festival-day turnover after the parade was moved out of the main street, where it was originally held.

The handful of local pubs and motels in Bundanoon receive the bulk of the economic benefits, with most guest rooms being booked out more than a year in advance. Some visitors stay overnight elsewhere in the region to avoid travelling long distances in the early morning or evening. One hotel bar even estimated that its turnover was 10–20 times higher during the festival. However, there are insufficient incentives for local businesses – particularly hotels – to expand their facilities or workforces on a more permanent basis. The concentrated – and highly temporal – cash impacts suggest that the festival makes a fairly small direct contribution to the local economy. Some benefits spill over to the wider Southern Highlands region, although these are also limited. Very few visitors had visited other attractions or undertaken other activities in either Bundanoon or the region. Many who stayed at night in the area stayed with friends or relatives.

However, the festival also brings non-cash benefits. These include the development of non-financial exchanges and relationships between economic actors at both local and wider scales, the sharing of resources in non-market ways, the donation of labour by individuals through volunteering, and contributions to tourism in the Southern Highlands region. In addition, the festival has partly shaped Bundanoon's place identity and put it 'on the map', potentially attracting longer-term tourism and in-migration as festival attendees return to the town.

Most brochures and media representations of Bundanoon make at least one reference to the festival or use distinctly 'Scottish' imagery, despite Bundanoon having little association with Scotland other than the festival. It has no significant history of Scottish immigration, yet it has adopted a Scottish identity that dominates the other features of the town when the festival is on. This is reinforced by the many Scottish-themed displays placed in local shop windows in support of the event. Over time, the festival has become a local attraction that brings in tourists and marks Bundanoon as 'Scottish'. This parallels the way that Celtic monuments and festivals have been used to develop a distinctive place identity and attract tourists to Glen Innes, NSW (Connell & Rugendyke, 2010).

In these indirect ways, *Brigadoon* has shaped Bundanoon's place identity and sense of community. Although many residents are not as drawn to the festival as they once were, and some leave the town during the festival to escape the mayhem, the festival has still become a local source of pride that both promotes the town to outsiders and brings together local community groups and their members. It is a successful example of a rural festival that has become the dominant source of imagery for tourism promotion in its host town, while also being a significant event on the local social calendar. In Putnam's (2000) terms, it creates 'social capital,' the goodwill 'glue' that binds together members of a community through networks of reciprocity and trust.

Experiencing and Consuming Scottishness

There are further dimensions to *Brigadoon*'s success than its local social and economic outcomes. We now turn to the reasons why so many people travel for several hours to crowd onto an oval filled with tartans, bagpipes and kilts. Through conversations with festival visitors, organisers, stallholders and local businesses over three years we investigated why *Brigadoon* is so popular, and what draws people to attend. Many visitors are not able to say exactly why they are there; they refer to the overall atmosphere and their enjoyment of bagpipe music, watching sporting activities or simply having a day out with family and friends. For some true motivations may go deeper.

Brigadoon is a distinctive event that attracts a large yet diverse crowd. Attendance is a way to spend time with family and friends and do something one cannot usually do on a typical weekend. For many of the participants in activities, attendance is an important part of their involvement in a band, dancing troupe or clan community, and a way to meet others with similar interests. The festival also provides opportunities to dress up and act 'Scottish', to watch distinctly Scottish events and listen to particular types of music, and to act out forms of cultural pride. Yet visitors' motivations vary in intensity. For some, attendance is simply about having fun; for others it involves a curiosity about their family history, and for still others it is an annual ritual through which Scottish-Australian identities are affirmed and performed. These themes emerged when respondents talked about why they attended, even if not all visitors could precisely identify what drew them to *Brigadoon:*

> It's back to the way things used to be in Scotland ... Just people having a good day. (Festival visitor 2006)

[I come] to see Scottish culture, eat haggis, toss a caber, learn about heraldry and clans and drink scotch with friends and family. (Festival visitor 2008)

[We attend] because we love the Scottish tradition ... [and] we just love the music. (Festival visitor 2008)

The majority of visitors, unsurprisingly, either identify as having some Scottish ancestry or personal connection to Scottish culture, or attend with others who do. Seeing the festival, participating in the various activities or engaging with the displays and 'genealogical discourses' of clan tents are ways of expressing curiosity for family histories or expressing an interest in Scottish culture. In some cases, it is about developing a deeper and more spiritual attachment to the idea of Scotland that has become a basis for personal identity and which generates a sense of belonging.

The size and popularity of the festival is largely in line with the history of substantial migration from Scotland to Australia, often many decades (or centuries) ago. More broadly, family history research and tracing genealogy have gained widespread popularity in the main settler countries of Australia, New Zealand, the United States and Canada (Basu, 2007; Nash, 2002; Timothy & Guelke, 2008). In Australia, around 1.5 million people identified as having Scottish ancestry in 2006, up from about 500,000 in 2001 (ABS, 2006), indicating a greater propensity to identify as being of a specific non-Australian cultural background. Most respondents at *Brigadoon* were only part-Scottish, and only a handful knew relatives in Scotland. An even smaller number were actually born in Scotland. It is likely that most Scottish-background visitors see themselves as Australian, but at occasions such as the festival they are Scottish-Australian. Like descendents of Scots elsewhere, they engage with this ethnic background through participation in festivals (Ray, 2005), tourism to the Scottish Highlands (Basu, 2007) or the consumption of music, food, dance and designs perceived to be inherently 'Scottish'.

Indeed, for some Australians of Scottish ancestry, Scottishness has been adopted as a distinct ethnic identity that distinguishes them from other white ('Anglo') Australians, and as a way to situate themselves in multicultural Australia:

We're here to stimulate the Scottish culture, and preserve it, because it's very much under threat in today's multicultural society. (Clan official 2008)

Scottishness might also be used to distance oneself from a history of English colonisation, convict shipments, indigenous dispossession and associated narratives. By investigating known links to Scotland, or uncovering new identifications, individuals and families can seek out seemingly more stable identities. By searching for cultural roots, a connection to a seemingly authentic ancestral place and culture is also established:

> [Genealogy] does seem to give a sense of identity ... There is this need for a sense of 'who the hell am I?'... If you've got a family history that you can look to ... if you see yourself as part of this unbroken line, it does give you this sense of 'who I am,' this sense of self-worth. (Clan official 2008)

> It's a very personal thing. In a society that's subjected to such rapid change ... people are looking for something that speaks to them of security in tradition and heritage. There is something satisfying about being able to say, 'Hey, this is who I am, because this is where I've come from in a family line' ... It does seem to give a sense of identity. (Clan official 2008)

Other research has found that tracing genealogies and engaging with ancestries (e.g. through ethnic organisations, travel, music or by displaying certain objects in the home) might serve as some sort of psychological buffer against a sense of constant change in society and the disruption of identities, especially in a context of multiculturalism and globalisation (Basu, 2007; Nash, 2002). Investigating one's cultural heritage and subscribing to the myths and stories of a distant 'homeland' and culture can also provide a sense of ontological security (Baldassar, 2001; Basu, 2007; Ruting, 2008). Moreover, involvement in Scottish groups and activities (clans, bands, dance troupes), adopting particular cultural practices or forms of clothing (tartans), and visiting Scotland itself are ways of connecting with the birthplaces of ancestors and can elicit a sense of belonging or inherent, irrevocable attachment to both place and history.

The motivations of many *Brigadoon* attendees can be situated in this context to varying degrees. Those who were engaging with the clan tents (Figure 16.3) often expressed a desire to research their family history and ancestral connections. Some individuals were highly dedicated to this task. When asked why they were doing so, many said that they had reached an age (typically around 40–50 years old) where they suddenly felt a need to research their family history and cultural background. This may have followed a significant event (such as a death in the family) or the onset of a feeling of mortality that compels them to find out more

Figure 16.3 A lone piper performs in front of the tents of Clans Fraser, Macleay and Macdonald (Photo: Brad Ruting)

about their family history – and pass this onto the next generation – while they are still physically able to do so:

> It's probably only when people get to their 50s or … they have the time to spend on this sort of thing … Young people are so forward looking, they don't sort of see the day when the Grim Reaper might be approaching … I wish I'd asked … my father or mother. But often because you don't have the background, not only knowing the answers, but often you don't even know the questions to ask. (Festival visitor 2008)

> We get a lot of people that have found out they're Scottish, they've got their name either back in their female line or their male line, and want to know more. And I find it's usually after a death in the family where they've forgotten to ask the person before about their father or their grandfather. (Clan official 2008)

Genealogy is invoked at *Brigadoon* through the creation of a sense of community of Scots and a sense of common descent from a spatially and temporally distant land. The festival brings together people from many different places, with different life histories and experiences. Nevertheless, most see themselves as sharing a common Scottish ancestry with other participants, and there is a common emotional bond to Scotland.

Clan societies play a significant role. Clans are organised groups claiming a common family name or ancestor. Essentially, they are networks of people with a shared (but often distant) family heritage, which one might join after tracing one's ancestry. Members are often spread across Australia and the world, reflecting the dispersions of migrants in past generations. For some members, clan history becomes personal history, with the clan taking on the status of an 'extended family' (Basu, 2007).

Clan groups also have their own websites, newsletters, chieftains, heraldry, tartan designs and events, including dinners, dances and gatherings. *Brigadoon* is marketed as a 'gathering of the clans' and brings these communities together from across Australia. There are over 30 clan tents at the festival, each displaying emblems and banners, and providing reference books to help visitors identify the clan to which they belong. The clan tents at *Brigadoon* become sites for tracing roots, venues of discovery and sometimes departure points for journeys into family history. Many people, of various ages, approached people staffing the clan tents to find out more about either the clan or their own family history:

> The ones walking around the clan tents, they're looking for something. (Clan official 2007)

> I think they're looking for roots. They're looking for kinfolk. You know, that sort of thing. (Clan official 2008)

Clans and other Scottish-Australian social groups also serve as both real and imagined communities. They are real communities of individuals who meet each other regularly and are involved with various activities. At the same time, most are also 'imagined communities' on a different scale, linked to other clan members living elsewhere in Australia, or even transnationally connected to Scotland and other places in the Scottish 'diaspora'. These communities are imagined since they consist of many others to which one is somehow related, but who are not known personally (Anderson, 1991). Aided by the internet and newsletters, they go beyond the local geographies of gatherings and festivals, allowing dispersed people to discover or reinforce their ethnic, cultural and even familial identities. They are part ancient extended family, part social network and part genealogical society – imagined but organised communities.

At *Brigadoon*, the sense of community and common purpose is achieved through participants seeing themselves as Scottish as part of a loose network of Scots spread across the world. As with many who have traced their ancestry, a Scottish background can become an anchor to something bigger and older than oneself. As genealogical research shapes new and existing identities, this network becomes an 'imagined diaspora'. In the Scottish case, particular stories of emigration and dispersal have sometimes been more sensationalised or emotive than the historical record would suggest (Basu, 2007). Notably, being a member of this 'imagined diaspora' is based upon self-identification rather than actually being born in the Scottish Highlands.

Discourses of family and clan history, and the meaning of Scottish traditions, are prominent at *Brigadoon*. Some respondents take these traditions with great seriousness:

> Some of the clans are more Scottish than the Scots. They are more low-key in Scotland, and we are more proud of our heritage here. It's an Australian thing. (Festival visitor 2007)

However, some clan officials and festival visitors expressed concerns about the continuity of social groups that have formed around notions of Scotland and Scottishness. Some felt a need to pass rituals and knowledge onto children, but were frustrated when their children did not show the same enthusiasm. Nevertheless, many believed that as their children get older, they will eventually become involved:

> I plan to go back to live [in Scotland] in five years or so … [And I'll] take my children, and you know, heritage I think is important, having a sense of belonging and knowing where you came from. I think that's very important to instil in children. (Festival visitor 2008)

The sense of community among festival visitors extends further than just the clan societies, which can be exclusionary as membership is based on proving genealogical descent. Indeed, Scottishness is also performed in other ways. The festival brings together for its 'Highland gathering' many bagpipe bands, for which *Brigadoon* is a key annual event. Indeed, the presence of bands in the parade, around the oval and in the 'massed bands' display at the closing of each festival provides visitors with a sensual and engaging representation of Scottish traditions. Yet while the militaristic parade and massed bands display a fierce sense of cultural pride, the amicable nature of all events and the diversity of stalls create an inclusive and friendly atmosphere. Within the network of pipe bands and clans, and also on the oval itself, a different type of community is created at *Brigadoon*, one which brings together participants, visitors and volunteers.

The sights, sounds and smells of the festival triggered a longing for a distant Scotland in some visitors – whether as a place of childhood or an imagined place of ancestral origins. A small number expressed a sense of nostalgia, although this was partly a longing for family members who had passed away or for respondents' own childhoods, rather than for Scotland itself. The sensory experiences of watching or participating in activities elicited feelings of nostalgia and cultural pride, recalled memories of childhood and family, or stirred the imagination and a passion for being Scottish. Many respondents spoke of their families and childhoods without being prompted; others were entranced by the bagpipes:

> Bagpipes are very emotional ... The sound sends shivers up the back of your spine. (Festival visitor 2007)
>
> The bagpipes have their peculiar sound. I find some of the tunes very moving. *The Flowers of Scotland*, I find it ... [lost for words]. (Festival visitor 2008)

The festival created a miniaturised and essentialised version of the Scottish Highlands in which visitors could immerse themselves. However, the feelings of nostalgia were usually not simply a melancholy sense of loss or displacement, as nostalgia is often defined, but rather a feeling of reflection and appreciation: a way to put the instabilities of the present into perspective through seeking familiarity and renewal in the real or imagined past (Pickering & Keightley, 2006).

However, some elements of the festival were seen as detrimental to the Scottish authenticity of the event by a handful of visitors. These included 'ethnic' foods (stands serving Chinese and Turkish cuisines), a significant number of stalls selling no or few Scottish items, and a sense that as the festival had grown it was compromised by needing to attract sponsorship. Nevertheless, many visitors explicitly said that they appreciated the mixture of 'Scottishness' with the rural Australian setting and an overall sense of (Australian) friendliness. Scottish culture was mostly represented in ways that entertain, or evoke emotional responses in, visitors. Particular elements of Scottishness were idealised and overemphasised, such as bagpipes, clan symbols, kilts and particular sports – most of these were never commonplace in the lives of Scots in Scotland and many are 'invented' traditions (Trevor-Roper, 1983), relating at best only to the Scottish Highlands rather than the cities and 'lowlands' where most of the population has always lived (Basu, 2007).

Indeed, respondents' perceptions of the authenticity of the festival – in the sense of the historical 'accuracy' of the activities – was often related to how strongly Scottish they felt. Those who were aware of their Scottish ancestry but had not actively researched it may come to identify with these symbols or enjoy them at such events, while people who intensively research their genealogies may come to see these as 'cliché' and 'kitsch,' preferring instead to identify with specific clan histories and particular Scottish places and landscapes. Nevertheless, most attendees thought that the representations of Scottishness at *Brigadoon* were largely authentic, and this was reinforced by the emotive reactions of some visitors to the sounds of bagpipes.

Indeed, what matters is not whether the practices and artefacts are 'authentic' in some historical sense, but the creation of a *sense of authenticity* at the festival. The notion that Scottishness is being portrayed 'properly' at *Brigadoon* is what elicits pride and cultural sentiments, while bringing diverse and dispersed groups of people together. The festival *feels* Scottish for most visitors and is sufficient to elicit sentiments of belonging, ancestry and cultural heritage. Indeed, perceptions of authenticity are arguably more determined by consistency with widely held images of Scottish culture and traditions rather than the historical accuracy of what is displayed. This is reinforced through the creation of a pastiche of cultural stereotypes at the festival, but in a manner that romanticises and privileges Scottish traditions and a sense of being descended from an ancient and noble culture. Hearing bagpipes, wearing tartans and eating haggis become ways to consolidate Scottish identities in Australia:

> They are trying to squeeze in as much as they can of what Scottish people do ... people get their money's worth. (Festival visitor 2006)

> We include [in the festival] any activity, but it must stay in the field of the Celtic. (Festival official 2006)

Several respondents were unconcerned about true 'authenticity' of the spectacle, and they simply wanted an enjoyable time with a playful (yet self-consciously uncritical) encounter with Scottishness. The organisers had in mind a fun and festive atmosphere that would attract large numbers of people, yet where visitors could interpret Scottish traditions however they wished. Indeed, not all visitors held emotive and ethno-culturally based sentiments; many were aware of having some Scottish ancestry but did not want to research it intensively; some others even considered the whole thing to be kitsch and 'just a bit of fun,' yet with some underlying importance:

> You keep an open mind about it. You remember that a lot of it is in a sense artificial. But look, it's a way of uniting people ... a sense of ancestry. (Festival visitor 2008)

On the other hand, for some non-Scots, the atmosphere was contagious:

> I'm disappointed that I don't have Scottish heritage ... I think because I feel a sense of connection here, but the history tells me that I shouldn't feel it. (Festival visitor 2008)

Conclusion

As *Brigadoon* shows, festivals can function as sites of identity and pride both within local communities and among the communities formed by their visitors. They can serve as focal points for local residents to come together and to promote their town to the wider world. Yet as grounded in place as festivals may seem, they can also transcend their host towns and provide a festive time and place for a significant number of people with a common interest to come together. Festivals are sites for articulating, negotiating and above all celebrating various identities and attachments.

In many ways *Brigadoon* is out-of-place in its Australian rural setting. It is a *sight* of Scottish cultural pride and a tourist attraction, and a *site* for the articulation of Scottishness and the strengthening of the dispersed communities that form clan societies. It is a ceremonial yet colourful celebration of Scottishness, a place where genealogies are traced, family histories are celebrated and a sense of common purpose emerges. Some visitors go to seek out their roots and family history; others want merely a fun day out with their family. An imagined collective identity of Scottishness is performed and sustained through this festival, even though 'Scottish' or even 'Scottish-Australian' may not be prominent identities for many attendees in their everyday lives. Some visitors have no Scottish ancestry whatsoever. Yet this festival allows Scottishness to come to the fore, for other differences to be put aside and a common connection – however tenuous or imagined for some – to be celebrated in a fun environment.

Brigadoon offers a temporal site to produce and consume Scottishness by playfully recreating a mystical and 'authentic' Scottish village in a small Australian rural town. Its direct cash benefits may be small and concentrated, but there are sizeable spillover benefits for the local community. The roles played by management, community groups and visitor expectations and perceptions are integral to the social and economic impacts of festivals. Those that are successful go beyond measurable outcomes and generate communities within which anyone can belong. At *Brigadoon*, one does not have to be Scottish to feel Scottish, if only fleetingly.

Acknowledgement

We thank Alaistair Saunders, of the *Brigadoon* organising committee, for his helpfulness and enthusiasm.

References

Anderson, B. (1991) *Imagined Communities: Reflections on the Origin and Spread of Nationalism*. Revised edition. London: Verso.
Australian Bureau of Statistics (ABS) (2006) Ancestry by country of birth of parents for time series. *2006 Census of Population and Housing*. Catalogue 2068.0. Canberra, Australia: ABS.
Baldassar, L. (2001) *Visits Home: Migration Experiences between Italy and Australia*. Melbourne: Melbourne University Press.
Basu, P. (2007) *Highland Homecomings: Genealogy and Heritage Tourism in the Scottish Diaspora*. Abingdon: Routledge.
Chhabra, D., Healy, R. and Sills, E. (2003) Staged authenticity and heritage tourism. *Annals of Tourism Research* 30, 702–719.
Connell, J. and Rugendyke, B. (2010) Creating an authentic tourist site? The Australian standing stones, Glen Innes. *Australian Geographer* 41, 87–100.
Nash, C. (2002) Genealogical identities. *Environment and Planning D: Society and Space* 20, 27–52.
Putnam, R. (2000) *Bowling Alone: The Collapse and Revival of American Community*. New York: Simon & Schuster.
Pickering, M. and Keightley, E. (2006) The modalities of nostalgia. *Current Sociology* 54, 919–941.
Ray, C. (2005) Transatlantic Scots and ethnicity. In C. Ray (ed.) *Transatlantic Scots* (pp. 21–47). Tuscaloosa: University of Alabama Press.
Ruting, B. (2008) Travel to the Old Country: Transnational Engagements and the Estonian diaspora. Honours thesis, The University of Sydney. Online document: On WWW at http://www.sites.google.com/site/bradruting/BradRuting-TraveltotheOldCountry.pdf. Accessed 15.05.10.
Timothy, D.J. and Guelke, J.K. (eds) (2008) *Geography and Genealogy: Locating Personal Pasts*. Aldershot: Ashgate.
Trevor-Roper, H. (1983) The invention of tradition: The Highland tradition of Scotland. In E. Hobsbawm and T. Ranger (eds) *The Invention of Tradition* (pp. 15–41). Cambridge: Cambridge University Press.

Chapter 17
What is Wangaratta to Jazz? The (Re)creation of Place, Music and Community at the Wangaratta Jazz Festival

R. CURTIS

> You get a fantastic beautiful confusion of inner city bohos and farmers and freaks and drunkards and yobbos and intellectuals and international super stars ... it is the master stroke that stirs the fires of creativity and is just downright more interesting. Cities are boring on their own, as are rural places. But mix 'em up and you get the Wangaratta Festival which is testimony to the crucible appeal.
> McAll, 2007

Introduction

Popular music festivals are valued for the thrill of live performance, their ability to attract tourists, economic growth and development and express social and political statements. The growth of popular music festivals around the world since the late 1960s has produced not only music but also place, space and identities. Music festivals provide places of spectacle and unique experiences, with one-off performances, networks of performers and audiences connected to particular musical genres. The festival space itself creates a dialogue of musical communication and emotional and affective responses, and the musical elements of festivals – the make-up of bands, musicians, performances, tunes and approach – are equally important. Despite music festivals being long established, there have been remarkably few studies of jazz festivals (cf. Formica & Uysal, 1996; Saleh & Ryan, 1993) and none at all in Australasia. This chapter considers how Wangaratta, an otherwise little-known country town in northern Victoria, has become a special place for jazz in Australia through the Wangaratta Jazz Festival, held there for 20 years.

This chapter explores what Wangaratta means to jazz in Australia by emphasising the relationships between music, place and people at the Wangaratta Jazz Festival. Jazz has become an important medium for nourishing social interaction between locals in Wangaratta, but also between jazz players and audiences at the festival. The sound of jazz in Wangaratta has shaped people's perceptions and experience of the place and the place has influenced the way jazz is made, heard and played. The jazz sound that is encountered at the festival intimately connects Wangaratta with major cities in Australia and overseas. Insights from eminent Australian and international players from different generations – older, well-established musicians and younger, up-coming players – are used as a lens for understanding the interrelation of jazz music and Wangaratta. The chapter is based on the 2007 and 2009 festivals and uses a range of research methods including formal and informal interviews with musicians, festival organisers and local residents in Wangaratta.

Wangaratta: Creating Place and Jazz

In Australia, jazz festivals have become increasingly woven into place promotion, for example, through the explicit linking of jazz sounds to place imagery, such as jazz 'in the "vines" or "in the pines" or "on the rock" or in the "valley" or "at the farm"', a practice that has contributed to the enhanced perception of such places as elite spaces (Gibson & Connell, forthcoming). Rural towns have used their mountains and forests, beaches and cliffs and vineyards and cheeses to create a sense of distinctiveness and attract people to a particular location to experience jazz and other cultural activities. Whether or not a local 'scene' exists, places have captured the sound of jazz as an economic, cultural and social resource. Jazz has become a means by which places can be 'put on the map', transform themselves and enhance their reputation and identity as significant places for cultural activities. Wangaratta has built a reputation as a place conducive to jazz music. The synergy between jazz sounds and the physical place of Wangaratta has been an important factor in the festival's success, and the Wangaratta Jazz Festival has emphasised its physical attributes and how they work with music.

Indeed, jazz music is said to have some degree of synergy with the physical geography of Wangaratta and its surrounding region. The region is known for its spectacular scenery, mountainous geography, gourmet food and for being home to 35 of Australia's top wineries, and Wangaratta advertises itself as a place that offers a premier provincial lifestyle (Rural City of Wangaratta, 2009). In 2006, the Milawa Gourmet Region, part of the Wangaratta region, used the area's reputation as a cultural centre to attract tourists to their very own 'Beat the Winter Blues 'n' Jazz Weekend'. The event took place in local wineries, cafes, restaurants and country pubs and its success could be attributed in large part to return visitors from the Wangaratta Jazz Festival (Jackson, 2007). The Wangaratta Jazz Festival has been acknowledged as a Tourism Victoria 'Hallmark Event' and is Tourism Australia's 'most significant regional festival' (Rural City of Wangaratta, 2009). The region markets itself as a tourist hub from which to explore North East Victoria and Southern New South Wales (Rural City of Wangaratta, 2009). In 2006, while the Festival attracted crowds from Melbourne (35%), a substantial proportion of the festival-goers were from Wangaratta itself (16%) and the broader region (11.5%).

Wangaratta's appeal as an attractive location for a jazz festival was recognised by the festival's creators. In 1989, a group of local business people sought to establish the festival as a landmark cultural event which would attract tourists to the town and boost the local economy (Wangaratta Festival of Jazz, 2007). The decision was based on no preexisting interest in jazz music, but was born from a feasibility study. The primary motivation of the organisers was to establish a credible event that provided the youth of the city with musical and tourism-related experience, while also generating revenue for the local economy through increased tourism (Clare, 1999). Bob Dewar, involved in planning the first festival, argued that musical credibility was central to its creation:

> We just thought that Wangaratta, the whole area, needed a kick, an infusion of money, some event ... you know of course that North East Victoria is a food bowl. Quality wines, meats, vegetables. Good food, good wines, good music. It all seemed to go together. When we decided on a jazz festival my idea was to appeal to the musicians, not the public! That may sound funny, but I felt that if it received credibility from the musicians the people would follow. They followed the musicians. That was how it would gain a reputation ... we would have a festival of credibility. (Dewar cited in Clare, 1999: 38)

Wangaratta does not support a local jazz scene, yet it has become a place where jazz belongs. The festival creates a space that does not exist anywhere else in Australia. As the Artistic Director of the Sydney-based Jazzgroove Association, John Hibbard explained: 'It's amazing when you go to this town, suddenly it becomes jazz land for four days of the year' (Hibbard, 2009). The physical space of the festival encourages community and cultural interaction and emotional and physical responses to music. All performance spaces – including the performing arts centre, blues marquee, local hotels and a free stage on Reid Street – are within close walking distance. At the heart of the festival is Reid Street, a community space where festival-goers and locals can enjoy local, regional and international food and wine in an open-air street mall. Being in Wangaratta for the festival is a musical, geographical and socially liberating experience for players and festival-goers. According to pianist Mike Nock:

> [a] lot of the greatest music festivals all over the world are in little out-of-the way places. You have to go somewhere to get there. You kind of get sequestered for the time in a different world. If you go from listening to one gig to another in the city it does not have the same intensity. Every year it is always great and the bar is constantly being raised. You will hear things that you will not hear anywhere else. (Nock, 2007)

As pianist Barney McAll explains:

> Wangaratta's location is a serendipitous master stroke on Adrian Jackson's [the festival's Artistic Director] part. Being between Melbourne and Sydney makes it perfect as a neutral place and the rural setting means you can actually play in a more open way. You are less laden with city life and complications and in a small way I feel this is apparent in the music played at Wangaratta. (McAll, 2007)

The performance spaces are intimate venues for listening to and hearing music, and for stimulating the community in a manner that has few urban parallels.

The Significance of Wangaratta as a Social, Cultural and Economic Landscape

Wangaratta has fostered a sense of place and belonging for jazz and a physical space for encounters with sounds, people and players through its festival. Music plays a significant role in facilitating notions of community and collective identity, grounded in physically demarcated urban and rural spaces (Gibson & Connell, 2005; Whiteley *et al.*, 2004). According to pianist Mike Nock, music is only half of jazz, the other half being the social and cultural context in which jazz takes place (Nock, 2007). Wangaratta brings these elements together.

Festival-goers are not only an audience for the music, but also to the local social and cultural environment (people, venues, cafes, wineries, bars and institutions). In 2009, the festival was host to a range of cultural activities from food and wine to art and religious gatherings. The Holy Trinity

Cathedral Wangaratta held a traditional 'jazz mass' on the Sunday, with live music, and Reid Street held a service of 'jazz praise' on the community stage. Several local wineries, including Auldstone Cellars and John Gehrig Wines, hosted jazz gigs by festival artists. Several exhibitions were held including the jazz-related displays of 'jazzART' and 'Visions of Jazz', the 36th Wangaratta Art Show and the 39th Wangaratta Artists Society exhibition of pottery, woodwork, silk scarves, art and craft. Reid Street was host to a multicultural mix of local and exotic food, local wine and music, and in 2009, it included Indian, Thai and Central American food stalls. According to Kate Green, one of the 2009 organisers of Reid Street:

> While the wine is from the local area, we've a few new food vendors from wider afield to tempt hungry festival-goers ... it [Reid Street] has a wonderful vibe and atmosphere. It is no wonder it continues to attract locals and visitors to enjoy its great mix of music, food and wine. (Green in Wangaratta Festival of Jazz Guide 2009: 19)

For local businesses and residents, the festival mobilises local community spirit, pride and identity. Since its initiation in 1989, it has been run primarily by volunteers: an honorary management board and hundreds of local residents who mobilise during the event, and for whom the festival is a way of helping the community. An inaugural Hall of Fame for festival volunteers was established in 2009. Perc Farrell, a volunteer for 19 years at the festival with his wife Bev, says that the reward for volunteering is watching the success of the festival: '[i]t's good to see so many locals and people from outside having a good time' (Farrell in Wangaratta Festival of Jazz guide 2009: 9).

From modest beginnings in 1990 as a two-day event, the Wangaratta Festival of Jazz has grown to be a four-day event of jazz and blues, showcasing more than 350 young and established Australian and international artists. The launch of a new multimillion-dollar performing arts centre in October 2009 as the festival's primary venue, replacing the old Town Hall that served as its previous centrepiece, is symbolic of the progress since its inception. The establishment of this new venue signifies the emergence of a new performing arts consciousness in the community and offers a gateway to community arts during non-festival periods.

The Wangaratta Festival of Jazz has had a significant local economic impact. In 1996, 51% of attendees were professionals (Wangaratta Festival of Jazz, 2007). Given the income-earning power of this group, jazz has become a significant vehicle for economic invigoration in rural towns in Australia (Gibson, 2007). In 2006, 50% of festival-goers spent more than $500 at the festival – with 16% of this group spending over $1000 – on accommodation, food, fuel, shopping, souvenirs and local produce, excluding festival passes and tickets. The Wangaratta festival, however, has not been central to the creation of a local cultural economy centred on jazz (in the sense of economic activity supporting the production and distribution of music), in stark contrast, for example, to Byron Bay in northern New South Wales, where its identity as a musical mecca and alternative lifestyle location was born out of the annual Blues & Roots Music Festival, accentuated by migrants moving there for its for lifestyle and amenity, creating a new market for cultural products (Gibson & Connell, 2003: 182). Outside the festival, jazz music is otherwise largely absent from Wangaratta. But this does not mean the festival and its location are any less significant. Co-Artistic Director of the Sydney-based Jazzgroove Association, Matt Ottingnon, believes that the festival's success is inextricably linked to the place of Wangaratta:

> The success of the Wangaratta Festival of Jazz is due to the fact it is in Wangaratta. Because it is a national jazz festival. If it was in Sydney or Melbourne, its successes would not be the same. (Ottingnon, 2009)

In other words, Wangaratta is unique and distinctly identified with the festival (Figure 17.1). Because Wangaratta is a small place, with a population of about 15,000, the festival absorbs more of the townsfolk than at any other time of the year, and has placed the town on the map.

Music and Belonging in Wangaratta

> I noticed coming into Wangaratta that there are signs up saying it is Australia's jazz capital. It just cannot be. Only for one week. But the amazing thing is there is probably more hard-core jazz going on there in that week than anywhere else, certainly regionally anywhere else in Australia, at any other time. (Hibbard, 2009)

While Wangaratta has taken jazz for its festival, it has not captured the sound of jazz as an everyday part of the lives of its local population. While the locals have grown in appreciation of jazz, this has not sustained a local scene. For John Hibbard, the festival is 'very much like a circus – it arrives and then leaves and no one takes up juggling' (Hibbard, 2009). It has however encouraged more music programs in local schools, but these have not focused explicitly on jazz music. Arguably, the festival has had a larger non-musical local impact through educating primary-school students on road safety through its sponsorship by the Transport Accident

Figure 17.1 Wangaratta, 'Australia's Jazz Capital' (Photo: Rebecca Curtis)

Commission. Festival organisers have previously attempted to shape local music culture through regular jazz performances in local pubs, but without success. A Technical and Further Education (TAFE) Institution program was established in a nearby regional town as a centre for music excellence in jazz, but was folded through lack of local interest. According to Festival Artistic Director, Adrian Jackson, 'In the beginning the locals were passionate about the festival. They were not in it for the music; they were in it for Wangaratta' (Jackson, 2007).

Wangaratta is instead a temporary neo-tribal meeting point for performers where they can talk together, listen to and hear sounds together, make sense of sounds together, perform together and drink and eat together (Hibbard, 2009; Ottingnon, 2009). It is a place that bonds players, rather as diasporic populations achieve meaning and significance through sharing an imagined spiritual homeland. The festival offers artists four days of the year when they can gain an injection of inspiration and form enduring social and musical relationships. According to Matt Ottingnon,

Wangaratta plays a special role in the regeneration of Australian jazz music because it 'is closer than New York' – a place considered by many players as the global jazz 'capital' (Ottingnon, 2009). For the winner of the 1997 National Jazz Awards, drummer Will Guthrie:

> [my] strongest memory of the festival (and the Australian scene in general) is one of mutual encouragement and support between musicians. Without this, and the passing on between generations of knowledge and experience, what would this music be? What would it mean? Not much me thinks. (Guthrie quoted in Wangaratta Festival of Jazz, 2010a)

The festival offers performers employment opportunities, learning and inspiration. Jazz players perform in many ensembles, musical styles and with many different musicians over the course of their careers. They are rarely paid salary, but are hired for an evening, a week or an international tour of several months duration (Monson, 1996). Jazz players thus rely on networks of playing relationships within and across scenes and artistic communities for their work. In this context, Wangaratta gives players what they need in a raw but supportive professional environment. Pianist Barney McAll explains that Wangaratta provides a place where musicians can catch up:

> Musicians are a strange lot and tend to isolate themselves in the sense that you might make a significant recording with some musos and then, never see them again, or you might go on the road all over the world and then when the band finishes, that is it. The 'family' is broken. However, at Wangaratta you can have a true catch up. (McAll, 2007)

Wangaratta has enabled those within the jazz community to feel that they belong at this place and this event, capturing social and artistic relationships that cement the event's reputation. As Saxophonist Julian Wilson describes:

> [e]very Wangaratta Jazz Festival is a truly memorable experience. 1994 was the first time I attended ... I spent Monday relaxing on the lawn at Brown Brothers watching my old mates (Eugene and Chris) play with Brownie's quintet. The day before I'd heard NUDE play a blistering early morning set. I was surprised to see people in the audience that I thought would be more comfortable listening to Brownie's band, grey beards swinging wildly. That must be what I love most about Wang. The fact that all the camps come together for one beautiful long weekend. The traditionalists, the conceptualists, the scientists and the heretics.

The prophets and the preachers. The protagonistas and the provocateurs. Musicians from all round the country get to share music, food, drinks and stories. This is where you can hear the best improvisers Australia has to offer, playing beyond their abilities...and then share a pizza with them! Wangaratta is to us what Christmas is to small kids. You spend the whole year looking forward to it and then don't want it to end. (Wilson quoted in Wangaratta. Festival of Jazz, 2010b)

The festival has built a reputation as the only jazz event in Australia with such a high level of musical credibility and integrity (Nock, 2007). The creation and implementation of the festival's musical identity by the Artistic Director, Adrian Jackson, is widely considered critical to its ongoing success and credibility. However, it was not a natural success; according to Jackson:

> Initially I didn't think it was going to work. I didn't think there would be a huge demand for modern and contemporary jazz in rural Victoria. In the first year we lost $50,000, although we received positive feedback. Some people believed in it and were willing to bank on its potential. (Jackson, 2007)

The musical success of Wangaratta is underpinned by three key events. The first is the National Jazz Awards, an annual competition between young musicians in a given instrument. These awards distinguish Wangaratta from other jazz festivals since this is the most important and prestigious jazz competition in Australia. The awards also provide a valuable opportunity for young musicians to meet, share ideas and receive exposure and recognition nationally. For saxophonist Julian Wilson, winning the National Jazz Awards in 1994 was important for his career:

> [o]n a professional level, the opportunities the festival has provided me have been invaluable. Winning the Award at such a young age was an amazing boost for my career, one that I've often reflected I was not fully prepared for, but an experience that pushed me forward, and continues to inspire me. (Wilson quoted in Wangaratta Festival of Jazz, 2010b)

Second is the National Jazz Writing Competition, which showcases writing about or inspired by the music. Third is a series of youth workshops designed primarily for secondary-level students and other visitors in town for the festival. These events, aimed at young musicians and writers, form a critical structure through which the exchange of musical ideas takes place. The involvement of mainstream media, such as Australian Broadcasting Commission (ABC) radio broadcasts, also plays an important role in the credibility and attractiveness of the festival.

Listening and Hearing Sounds at Wangaratta

Wangaratta is a festival of musical excellence. It is not a festival of easy listening, and the sounds of Wangaratta may be unfamiliar in a town where jazz otherwise has no presence. There is a distinction between those who 'hear' the music (understand and comprehend it) and those who 'listen' to it (not necessarily tied to interpretation) (Simpson, 2009: 2563). The festival is a space where festival-goers experience players who do not have to 'sell out' or 'water down' their music (Nock, 2007). According to pianist Mike Nock, at Wangaratta there is a pressure to play at your best:

> To me, I always have felt, ever since the beginning of Wangaratta, and I have been there since the beginning, that basically it is a place that demands that you give your best without any commercial considerations. That's how I have treated it. Not everyone would of course. It's a part of a contract you enter into with the audience. I feel totally that it is a space where musicians can be true to themselves and the music. (Nock, 2007)

Pianist Barney McAll said that at Wangaratta the musical experience is of high quality:

> Adrian Jackson has bridged the gap so that audiences can be happy with what they experience and the musicians are top notch and never compromise their sound. The wide variety, yet high quality, of musicians is key here. (McAll, 2007)

Festival-goers experience an uninterrupted and intimate listening and hearing encounter with improvised music, becoming enveloped by sound. Musical performance is a way of 'articulating the subtle complexities of emotion that language cannot access' (Smith, 2000: 631). At Wangaratta, the emotional power of music is intensified in jazz. The festival challenges audiences to expand their musical horizons by providing a forum for experimentation and improvisation. As Mike Nock explains the way he approaches music: 'I see myself as a painter. I don't start out with chords. I use colours and textures and let the music write itself, because it always tells you where it wants to go' (Nock, 2007). The shared temporality and space of people's hearing and listening experiences at Wangaratta contribute to a sense of connectivity for those present.

The experience of a performance at the festival does not exist solely in the performance itself but in a wider range of settings, from composing sessions to rehearsals (Wood *et al.*, 2007). The sounds at Wangaratta are experienced in performance spaces and also on the street, at a pub or in a café. The creation of musical phenomena is actively achieved through social practice, powerfully informed by the situation, the participant's goals in the event and a potentially endless range of larger cultural contexts (Berger, 1999). Listening to any song on one night of the festival enables a focus on the formal and emotional aspects of the song, the stage moves of the musicians, the taste of beer and other aural or visual elements of the event. The participant's comprehension and enjoyment of such experiences is influenced by their purposes in attending the event (the desire to get out of the house, to hear intricate and diverse music, to support the local festival) and their perceptual skills (listening skills, knowledge of music). These in turn are informed by past musical experience (years of listening to jazz, an interest in the Australian scene, music classes) and non-musical experiences (local environment, a good day), as pianist Barney McAll describes:

> Wangaratta is a place where the psychological forecast can be told. Musicians travel around a lot and when they congregate and play they are describing their travels and experiences in abstract terms. That is why audiences often close their eyes at Jazz concerts. They are not as interested in the visual as they are in the story being told. The global update. The local issues in musical form. The bush poetry. The music at Wangaratta and the music I play is a description of life. Often musicians get caught up in rehearsing the potent music of the 60s in the United States but I feel many of the Australian groups are finding a new original voice. Adrian Jackson has picked up on this and presented it as jazz. The leather elbows and jazz stiffs can go and enjoy it too. (McAll, 2007)

The emotional and social experience of the festival is both 'here and now' and simultaneously within other places and times. It plays a crucial role in facilitating diverse musical experiences which keeps Australian jazz fresh and alive. For jazz musicians returning to their city homes, the experience of Wangaratta lives on through performances in local settings which have been influenced by the sharing of music at the festival. As Mike Nock explains, the Wangaratta experience inevitably leads to the formation of new bands and assists musicians hear each other, and in turn, get gigs.

> Two of my most recent records have been directly as a result of Wangaratta. Adrian [the festival's Artistic Director] put that together. Dave Lieberman [a famous saxophone player] and I had not played for twenty five years and because Adrian brought us together we recorded and won an ARIA Award. It was particularly nice because it

was done with no planning – it was spontaneous. We just played, we talked about a few things, and did what jazz musicians are supposed to do. And my new record is a larger ensemble that Adrian first put together. It is actually a combination of Melbourne and Sydney bands called the Mike Nock project. It is the cream of the younger players and it formed totally out of Wangaratta. (Nock, 2007)

During the festival Wangaratta thus becomes a temporary jazz scene. The festival synthesises locally acquired musical knowledge from different urban scenes in one place at one time. It is not a scene embedded in place, but an imagined scene that transcends place on one level (a collection of scenes), but relies on it on another (the place of Wangaratta). Wangaratta acts as an incubator for new talent and connects Australian jazz to international playing styles and scenes from Europe, Asia and the Americas through international acts. The decline of jazz in Australia and the closing down of many venues in Australia's cities has meant that Wangaratta has become an even more important place for jazz musicians in Australia. It has also helped spread the language of jazz with hundreds more people in Australia listening to live broadcasts of the festival. As Barney McAll describes:

> It is just deeply about music. It is the best festival to play at for this reason. There is no sense of compromise; of sucking up to corporations; it is just legit and I think people register and appreciate this fact. I would just say that Wangaratta becomes a barometer for me in a sense. I have played at the festival every two years since its inception and I get a vivid gauge of how my music has changed every time I play. I have also created more and more fans through playing Wangaratta and Adrian has been very supportive of all my whacked-out ensemble ideas. There is nothing that makes me prouder than playing the Town Hall when it's full of people who really enjoy my music and its feels like a net below a tightrope that I can never injure myself on. (McAll, 2007)

Conclusion

Jazz music is not indigenous to Wangaratta, yet Wangaratta is the lifeblood of Australian jazz music. The Wangaratta Jazz Festival has become the means for an otherwise little-known country town to become a special place to jazz. While it is not a capital for year-round jazz in Australia, the town acts as a kind of Mecca for jazz. Jazz music is re-routed from cities to the country for four days of peace and music. The Festival has become an improvised sound and lived experience for four days of the year. The festival's impact lives on beyond the physical boundaries of the festival through new social connections, recordings and ensembles which are born out of it. It connects city and rural populations in Australia and internationally through a sense of belonging in place and at the event.

The Wangaratta Jazz Festival is a festival of artistic excellence which fosters highly original developments in Australian jazz. There is no other jazz festival in Australia that achieves the level of musical excellence at Wangaratta, according to leaders in the Australian jazz community. For another place in Australia to have the success of Wangaratta, it would need to have a range of essential elements: a well-known food and wine culture; ideal location; welcoming local community; centrally located performance venues; and musical excellence to the standard of a national jazz festival. The Wangaratta festival is important to locals, players and festival-goers for different but related reasons. For locals, as the largest event on the calendar, the festival plays a valuable role in community building and economic invigoration. For players, the festival is a space where people are free to perform, talk and 'do what jazz musicians do'. For festival-goers, it is a place where people can experience sounds and emotion in an intense, uninterrupted and physically appealing environment. Finally, the Festival enables a better understanding of the interactions between festivals and jazz, and between a small country town and a festival that was much sought after, despite its wholly metropolitan credentials.

References

Berger, H. (1999) *Metal, Rock, and Jazz: Perception and the Phenomenology of Musical Experience*. London: University Press of New England.

Clare, J. (1999) *Why Wangaratta? The Phenomenon of the Wangaratta Festival of Jazz*. Wangaratta: Wangaratta Festival of Jazz.

Formica, S. and Uysal, Y. (1996) A market segmentation of festival visitors: Umbria jazz festival in Italy. *Festival Management and Event Tourism* 3, 175–182.

Gibson, C. (2007) Music festivals: Transformations in non-metropolitan places, and in creative work. *Media International Australia incorporating Culture and Policy* 123 (May), 65–81.

Gibson, C. and Connell, J. (2003) Bongo fury: Tourism, music and cultural economy at Byron Bay, Australia. *Tijdschrift voor Economische en Sociale Geografie* 93, 164–187.

Gibson, C. and Connell, J. (2005) *Music and Tourism: On the Road Again*. Clevedon: Channel View Publications.

Gibson, C. and Connell, J. (forthcoming) *Music Festivals and Regional Development*. Aldershot: Ashgate.

Hibbard, J. (2009) Interviewed by Rebecca Curtis. Canberra, 2009.

Jackson, A. (2007) Interviewed by Rebecca Curtis. Melbourne, 2007.

McAll, B. (2007) Interviewed by Rebecca Curtis. New York, 2007.
Monson, I. (1996) *Saying Something: Jazz Improvisation and Interaction.* Chicago: University of Chicago Press.
Nock, M. (2007) Interviewed by Rebecca Curtis. Sydney, 2007.
Ottingnon, M. (2009) Interviewed by Rebecca Curtis. Canberra, 2009.
Rural City of Wangaratta (2009) Discover – Adventure, legends & indulgence. On WWW at http://www.wangaratta.vic.gov.au/CA256B5800826065/page/About+Our+Region?OpenDocument&1=101-About+Our+Region~&2=~&3=~. Accessed 23.04.10.
Saleh, F. and Ryan, C. (1993) Jazz and knitwear: Factors that attract tourists to festivals. *Tourism Management* 14, 289–297.
Smith, S. (2000) Performing the (sound) world. *Environment and Planning D: Society and Space* 18, 615–637.
Simpson, P. (2009) Falling on deaf ears: A postphenomenology of sonorous presence. *Environment and Planning A* 41, 2556–2575.
Wangaratta Festival of Jazz (2007) *Research Report: Patron Attendances.* Wangaratta: Wangaratta Festival of Jazz (2009) *Relax in Reid Street Jazz Festival Guide 20th Year.* Wangaratta: Wangaratta Jazz Festival, p. 19.
Wangaratta Festival of Jazz (2010a) 1997 Will Guthrie – drums. On WWW at http://www.wangaratta-jazz.org.au/cms-jazz-awards/1997-will-guthrie-drums.phps. Accessed 23.04.10.
Wangaratta Festival of Jazz (2010b) 1994 Julien Wilson – saxophone. On WWW at http://www.wangaratta-jazz.org.au/cms-jazz-awards/1994-julien-wilson.phps. Accessed 23.04.10.
Whiteley, S., Bennett, A. and Hawkins, S. (2004) *Music, Space and Place: Popular Music and Cultural Identity.* Aldershot: Ashgate.
Wood, N., Duffy, M. and Smith, S.J. (2007) The art of doing (geographies of) music. *Environment and Planning D: Society and Space* 25, 867–889.

Index

Aboriginal land rights 113, 115-6, 123
Aboriginal people 7, 12, 34, 39, 50, 97, 100, 109-21, 123-35, 140, 147, 158, 233-4
accommodation 96, 184-6, 269
Adelaide (SA) 37, 196-7, 198
– Adelaide Grand Prix 79
advertising 183; see also sponsorship
Africa 140
age 3, 8, 64, 182
agriculture 3, 10, 19, 26, 35, 64-5, 101, 156, 211, 265
– agricultural shows 8, 9, 10-11, 25-43, 65, 100-101, 136, 144
akubra 149, 251
Albion Park (NSW) 12
alcohol 3, 13, 48, 118, 252; see also beer, wine
Alice Springs (NT) 11, 39
Anzac Day 32
Arnhem Land (NT) 109-121
art 120, 156
artists 19, 48, 100, 111, 233, 284, 287
arts and crafts 117
arts festivals 48
AstroFest (Parkes) 98
audiences 48; see also festival goers
Australia Council for the Arts 40
Australian Open Tennis Tournament 84-8
authenticity 151, 249, 272, 277; see also strategic inauthenticity

'back to' festivals 33-34
bagpipes 48, 265, 267, 275-6
Bali 102
Ballarat (Victoria) 5, 7, 19, 33, 138
Ballina (NSW) 78, 80
Bannockburn (Victoria) 98
Barraba (NSW) 10
Bass Coast (Victoria) 6
Bathurst (NSW) 97
beer 3, 101
belonging 44-55, 116-120, 126-134, 251-60, 262, 271
Bendigo (Victoria) 33
Bermagui (NSW) 49-55, 98

bluegrass 7, 146
Blue Mountains (NSW) 97
blues 7, 22, 146
Boorowa (NSW) xvii, 11
Bourke (NSW) 100
branding 3, 10, 22, 177, 196, 206, 254
Brigadoon Scottish Festival, Bundanoon 75, 265-79
Brisbane (Queensland) 29
brochures 4
Broome (WA) 13, 102
Bulli (NSW) 11
Bundaberg (Queensland) 30
Bundanoon (NSW) 11, 75, 265-78
bunggul 114, 116-9
bush regeneration 98
business development 14, 70, 72, 82-4, 85, 88, 187-88
buskers 179, 250
Busselton Wildflower Exhibition (WA) 157-8
Byron Bay (NSW) 7, 8, 78, 80, 95, 103, 285

California xv
camp 214-7
Canada 37
Canberra 85, 88, 266
Canowindra (NSW) 11
capacity building 61
Cape York (Queensland) 123-34
capitalism 120, 123, 222
carnival xv, 5, 7, 100, 217
cars 4, 13, 22, 35, 177, 202
Casino (NSW) 11, 101
ChillOut Festival, Daylesford 4, 8, 9, 209-25
Chinese 139-40
Chittering Wildflower Festival 157
chocolate 102
Christmas festivals 194-207
church music 178
circus 17, 31, 35, 229-46
citizenship 30-32, 36, 38, 137-8, 141-4
class 10, 12, 26, 30, 142-7, 261
classical music 49-55, 98

Index

Coffs Harbour (NSW) 6
Collector (NSW) 10, 101
colonialism 109-10, 123-30, 139
commodification 120-1
community xvi, 8, 20-21, 25, 39, 41, 44, 46, 52-3, 65, 72, 110, 128-134, 138, 141-4, 155, 184, 188-90, 199-201, 249, 260, 274, 283-5, 292
– imagined 46, 55, 274-5, 278
conservation 96, 98, 99
Cooma (NSW) 36, 38-9
Coonabarabran (NSW) 11
cosmopolitanism 10, 36, 38, 126, 140, 198; see also multiculturalism
country music 7, 9, 10, 11, 16, 40, 136, 145-7, 151, 231
Country Women's Association (CWA) 142
cowboys xv, 133, 139, 177, 216, 231
creativity xv, 5, 11, 46, 100, 113, 131, 168, 189, 280
cultural capital 38, 101, 254-5, 262
cultural tourism 111, 112

dance 112, 116, 133, 256-7
Darwin (NT) 36, 39, 110
Daylesford (Victoria) 5, 7, 8, 209-225
decolonisation 134
Deniliquin (NSW) xvii, 11, 13, 17, 22, 75
Devonport (Tasmania) 21
didjeridu 52, 111, 119
doof festivals 97; see also rave
Dorrigo (NSW) 7
drought 20-21, 147, 175, 186, 243
drugs 5
Dubbo (NSW) 185, 187
Dusty, Slim 231

ecological footprint 94-6
economics of festivals xvii, 74-89
– business surveys 76-7, 80, 82-4
– employment impacts 10, 16, 19-20, 80, 84, 188, 287
– income generation 14, 16, 75, 151, 186, 243, 267, 284-5
– investment 16, 19, 63, 70, 71, 72, 75
– multipliers 17, 79, 86, 187, 188
– visitor expenditure surveys 76-7, 79-82, 186-7
ecotourism 160
Eden Whale Festival 99
Edinburgh, Scotland 15
education 111-2, 117, 126, 161, 166, 206
elites xvi
Elvis (Revival) Festival, Parkes 175-192

emotion 8, 12, 44, 47-55, 156, 164-6, 258, 263, 265, 274, 292
– emotional responses to music 49-54, 276, 277, 282, 289-90
empire 30-32, 142-3
employment 19-20, 64, 11, 13, 176, 188
– in festivals 19-20, 69
– in music 287
environment xvii, 92-104, 158, 160, 195, 201, 202, 204-5, 206, 268
ethnicity 34, 139-40, 249, 261, 270-78
Ettalong (NSW) 5

Family 229-246, 251, 271-3, 274-5
festival goers 14, 79-80, 95-6, 111-2, 140, 157,182-4, 218-221, 235-6, 248, 250-59, 270-4, 275, 280
Festival of the Snows (Cooma) 38-39
Festival of Voices (Hobart) 14
festivals
– definition 4-5
– economic impacts; see economics of festivals
– environmental impact 21; see also environment
– film festivals 151
– folk festivals 7, 11, 250-62
– food festivals 4, 11; see also food
– gay festivals 4, 5, 209-25
– indigenous festivals 100, 109-135
– music festivals 7-8, 10, 14, 48, 49-55, 136, 145-50, 248-63, 280-92
Florida 15
flowers 9, 13, 48, 97, 102, 136, 141-4, 155-169
folk music 11, 248-63
food 4, 22, 101-102, 212, 276, 281, 282, 284
Forbes (NSW) 185, 187, 192
Four Winds Festival (Bermagui) 49-55
fun 5, 12, 16, 21, 98, 175, 183, 184, 212, 217-8, 236, 248, 263, 270, 277-8

gardens 9, 97, 102
Garma festival (NT) 12, 100, 109-121
gay culture 4, 8, 225; see also lesbianism
gender 141-2, 144, 149, 165, 214-7
genealogy 272-4
Germany 140, 196, 203
Glastonbury, England xv
Glen Innes (NSW) 11, 269
globalisation 121, 151
Gloucester (NSW) 11, 61-73
Gold Rush Festival, Gympie 138-40
gospel music 178, 182
Graceland (Memphis) 178, 184

Gracelands (Parkes) 176, 177, 184
Grafton (NSW) 5
Great Britain; see United Kingdom
Gulkula (NT) 111, 116, 119
Gundagai (NSW) 11
Gunnedah (NSW) 11
Guyra (NSW) 10, 101
Gympie (Queensland) 136-152

handicrafts 284
Harrow (Victoria) 40
Hepburn Springs (Victoria); see Daylesford
heritage 6, 11, 14, 41, 49, 97, 138, 143, 196, 198, 202, 205, 206, 212, 249, 272, 274, 277
– and music 263
hip-hop 7
Hobart (Tasmania) 14, 97
home 242, 245, 254, 272
home hosting 186
Homeland Festival of Sacred Song and Dance (Dorrigo) 7
homophobia 222-4
homosexuality; see gay culture
Hopetoun (Victoria) 28
hotels 184-6

icons 11, 13, 114, 123, 139, 147, 149, 179, 190, 216, 217, 219
identity
– Australian 12-13, 30-2, 38, 109-21, 129-34, 209, 249
– local and regional xvi, 3, 12-13, 8, 65, 67, 109-21, 125-30, 136-52, 189-90
– national 48-9, 147-9
– performed 116-8, 241-3, 255-60
– personal 46, 194, 212-3, 214-5, 221, 240-3, 251-5, 258-9, 270-4
ideology 35, 249
image 151, 180, 188-90, 194-6, 209
imperialism 27, 31-2, 142-3
industry 64
infrastructure 17, 39, 93
innovation 68
Inverell (NSW) 98

Japan 97, 181
jazz 7, 10, 11, 21, 22, 146, 280-92

Kangaroo Valley Folk Festival 250-262
Kariong (NSW) 97
Katoomba (NSW) 97
Kenilworth Show (Queensland) 28
kitsch 177, 180-1, 183, 184, 191
Kurrajong (NSW) 11

Lake Macquarie (NSW) 6
land 110
landcare 98
landscape 52, 67, 92, 98, 101, 103, 141, 161-2, 164, 166, 169, 197-8, 204, 206, 283
Latrobe (Tasmania) 101-102
leadership 61-73
Lennox Head (NSW) 74-89
lesbianism 11
line dancing 3
Lismore (NSW) 85, 86, 88
Lithgow IronFest 10
Lobethal (South Australia) 194-207
local identity; see identity, local and regional
love 116, 142
lyrics 48

Macassans 116-7
Mapoon (Queensland) 12, 123-34
Mardi Gras (Sydney) 213, 214, 220
marketing 21, 138, 146, 147, 151-2, 195-6, 203, 213
markets 17, 158, 162, 187-8, 268
media xvi, 4, 137-8, 145, 177, 180-81, 274
Melbourne (Victoria) 7, 14, 85, 88, 96, 210, 211, 213, 230, 231
memory 125-130, 133, 137, 169, 203; see also nostalgia
Memphis 178, 180
metal (music) 11
Metropolis (Illinois) 191
Midnight Oil 123, 131
migration 6, 10, 50, 64-5, 127-8
mining 110-1, 114, 118, 123, 127, 137-9, 156
missions 126-7
Miss Showgirl contests 36
Mitchell (Queensland) 235-6
Moonta (South Australia) 33
Mount Elephant Festival (Victoria) 98-9
Mullewa Wildflower Show 157-8
multiculturalism 39, 47, 101-2, 111, 139-40, 233
multinational corporations 149
Murray River 9
museums 16
– Parkes 179, 190
musicians 117, 256, 281-91
Myrtleford (Victoria) 5

National University Games 85-8
nature 96-100, 160, 165-6
neo-tribes 8, 261-2, 286-8
Netherlands 140

Index 297

New York 287
New Zealand 27, 80, 157, 172, 231, 271
Newcastle (NSW) 6, 27, 29, 30, 31, 65, 66, 80
newspapers 34, 38, 145, 177, 190
Nimbin (NSW) 5
noise 12, 21-2, 53, 75, 93, 97; *see also* pollution
nomadism 17, 229-46, 257, 262
nostalgia 33-4, 128, 142-3, 146, 183, 184, 276; *see also* memory
Notting Hill (London) xv

Oberon (NSW) 97
Olympic Games xvii
opera 7, 99
Opera in the Outback (South Australia) 99
Opera in the Paddock (NSW) 98

Palm Creek Folk Festival, Queensland 251
Parkes (NSW) 12, 16-7, 22, 40, 75, 98, 175-92
performance xv, 286-91
Perth (WA) 37
Philippines 140
pilgrimage 34, 184
place
– place identity 12-13, 99, 155, 165-6, 168-9, 177, 195-203, 219-21, 241-3, 269-70, 281-5, 292
– place marketing 11-12
politics 7, 112-6, 121
pollution 93, 97; *see also* noise
population 85-8
Port Fairy (Victoria) 11, 249-50
Port Stephens Whale Festival (NSW) 99
post-colonialism 113
post-modernity 184
post-productivism 3, 9, 15, 100-2, 195
Presley, Elvis 12, 175-192
protest 7, 32, 56, 127, 213, 250, 260-1
pubs 82, 250

Queanbeyan (NSW) 5
Queen 216
Queensland 80, 126, 153
Quorn (SA) 34

railways 26-7, 31, 33-5, 178, 179, 192, 210
Ravensthorpe (WA) 155-169
reconciliation 112-3
regional development xvi, 9-12, 15, 25, 45
religion 109, 147, 196-7, 204
representation 44, 47, 49-55, 120, 166
retirement 6, 8, 9, 50
ritual 24, 52, 54, 97, 116, 120, 116, 122, 130, 138, 144-5, 157, 219, 303

Riverina (NSW, Victoria) 9-10
rock 'n' roll 6, 7, 145, 149, 178
Rockhampton (Queensland) 11
rodeo 11, 17
romanticism 34, 109, 195, 196, 198, 206-7, 237, 256, 277
Roswell (New Mexico) 191
Rotary Club 137, 143, 187, 214, 268
Royal Easter Show (Sydney) 29, 30
rurality 3, 12-13, 45, 47, 100, 146-9, 194-207, 209
– rural idyll 97, 194-207
Rusty Gromfest surf carnival (Lennox Head NSW) 17, 74, 77-89

Scarecrows 5, 11, 65-66
Scone (NSW) 11, 22
Scotland 48-9, 161, 191, 212, 265-79
sea change 9, 50, 152, 265
seasonality xv, 97
service industry 17, 19, 35, 64-5, 69, 80, 83-4, 89, 186-8, 211
Shakespeare Festival (Swan Hill) 36-8
Shinju Matsuri Festival (Broome) 13, 102
showgrounds 14, 27-9
Showmen's Guild of Australasia 28
Silvan (Victoria) 102
Singapore 102
Snowfest Festival, Gloucester 11
Snowy Mountains 6, 10, 38-9, 137
sociability 164
social capital 9, 270
social change 9, 35, 144, 151, 221-4, 265
South Australia 80
souvenirs 85, 96, 188
spas 210
Splendour in the Grass Festival, Byron Bay 8, 22, 95-6, 103
sponsorship 14-15, 181-2
sport 4. 5, 6, 14, 15, 21, 30-31, 74-5, 77-88, 100, 133, 213, 266
St Albans Folk Festival 250-62
stallholders 17, 187-8
strategic inauthenticity 191
sub-culture 8, 248, 255-63; *see also* neo-tribes
subsidies 40
Summit to the Sea cycling festival 8
surfing carnivals 74, 77-89
sustainability 7, 14, 93-6, 99, 192, 199-201, 205, 206
Swan Hill (Victoria) 36-8, 39
Switzerland 212
Sydney (NSW) 7, 14, 22, 29, 32, 96, 183, 220, 231, 244, 266

Tamworth (NSW) 7, 11, 40, 147, 149, 185, 186
Tasmania 14, 102
taxonomy 163-4
techno music 97
television 145, 232
Tenterfield (NSW) 9-10
Tesselaar Tulip Festival (Silvan) 102
Thompson, Jack 112, 114
Torquay (Victoria) 85
tourism 6-7, 34, 35, 66, 69, 72, 138, 157-8, 162, 210-1, 269
Toyota 149-50
tradition 32, 41, 47, 65, 100, 11-2, 116, 120, 121, 140, 151, 164, 198-9, 249, 255, 261, 263
– tradition, invented 176, 191, 277
Tumut (NSW) 5, 97
Tweed Valley (NSW) 7, 22, 99-100, 101
Tweed River Festival (NSW) 99-100

Uluru (NT) 39
uneven development xvi, 109, 133
United Kingdom 27, 29
United States 27, 29

Victor Harbor (SA) 192

Victoria 80
Victorian Seniors Festival 8
visitors; see festival goers
voluntarism 14, 20, 35, 68-70, 144, 159, 163, 213, 267, 269

Wagga Wagga (NSW) 6
Wandiligong Nut Festival 101
Wangaratta (Victoria) 280-92
Warwick (Queensland) 11, 184
whales 99
wine 101, 282, 284
Willochra Plain (South Australia) 34
Wintersun Festival 7-8
Wollongong (NSW) 6, 12, 80
Woolgoolga (NSW) 101
Woodford (Queensland) 7
World Cup of Athletics 85-8
World War II 35

Yaamma Festival (Bourke) 100
Yellow Gum Winter Flowering Festival (Bannockburn) 98
Yolngu 109-121
Yothu Yindi 110, 112-4, 117-8
youth 117, 138